Dents in the Ceiling
Tools Women & Allies Need to Breakthrough

Angel G. Henry

NEW Reads Publications | Jacksonville

Copyright © 2021 by Angel G. Henry

All rights reserved. No portion of this book may be reproduced, stored in a retrieval system, or transmitted in any form or by any means—electronic, mechanical, photocopy, recording, scanning, or other—except for brief quotations in critical reviews or articles, without the prior written permission of the publisher.

Disclaimer: All names and identifying details have been changed to protect the privacy of the individuals and the companies in which they work. Certain details were intentionally omitted to prevent the identification of individuals and their companies.

Published in the United States by NEW Reads Publications. NEW Reads Publications is a registered trademark of NEW Reads Publications, LLC in Jacksonville, FL.

newreadspub.com

Library of Congress Cataloging-in-Publication Data is available upon request.

ISBN 978-1-7357219-2-7 (paperback)
ISBN 978-1-7357219-3-4 (ebook)

Printed in the United States of America

Book design by Nikesha Elise Williams
Cover design by Gisette Gomez

First Edition: June 2021

For Levora (Oden) Wade, the ultimate example of Black women's resilience, and to Sophia Davis, a leader of the next generation of beautiful, powerful, intelligent lovers of science and technology, to whom I pass the baton.

Also for Avis for planting my seed, Kailei for watering it, and Dell, Dshawn, Moriah, and DJ for being the light, air, and soil in which I grow.

Contents

Foreword by Karen Arrington	vi
Introduction	12
Prologue	27
Chapter 1: Visibly Invisible	31
Chapter 2: Bullied at Work	63
Chapter 3: Deep Cuts	87
Chapter 4: Maid Service	129
Chapter 5: Emotional Misconduct	165
Chapter 6: Macro Hurts	197
Chapter 7: Put Me in Coach	245
Chapter 8: Resiliency Building	272
Epilogue	297
Resources	301
Acknowledgements	303
Bibliography	306

Foreword

Call it a book. Call it a devotional. Call it a 21st-century corporate playbook on how to get your groove back. No matter what you call it, *Dents in the Ceiling* is the holy grail of female empowerment for African American women climbing the ladders in STEM (accompanied by a big Iyanla-sized hug). Thirty real women, thirty life-changing stories across corporate America.

Fist up to Angel Henry, Empowerment Champion, for creating this movement to empower women across the globe and the next generation of female STEM leaders. Thank you to the brave and fearless women who removed their veil and shared their heart-warming, intimate, and inspirational stories on the pages to follow.

You will feel an emotional connection and community with these courageous and powerful women. You will laugh, cry and wave your hands in the air. The triumphant stories will catapult you from a private pity party to your next-level career. You will be moved and inspired to step into a career full of impact, adventure and radically up-level your self-belief, resilience, and commitment to your dreams.

Over the past 20 years, I've personally mentored and helped over 1,000 women step into their next-level careers, transcending hardship, racism, and extreme financial limitations. I know a thing or two about success. And if you think success is about luck, privilege, or having the right publicist, think again.

If you achieve any level of success, it won't be because things worked out. It will be because things went wrong. But you handled it. You handled it on a thousand different days in a thousand different ways, but reading the stories and clues here will help you learn how to overcome the pitfalls that we encounter all too often.

I was thirty-something when I had my first major career blow. My boss walked into my chic corner office and politely laid me off. "Let's take a walk" laid off. Naturally, I did what any alpha female would do. I fell apart. In the days and months that followed, there were times when all I could do was form a one-person prayer circle. Eventually, I pulled myself together. I went from fired to favored and created my own job, on my own terms. I haven't looked back. I learned how to become resilient!

If it were not for losing my dream job, I wouldn't know what being a seven-figure entrepreneur looks like. Life hasn't been easy, but through the hardships, I have mastered how to turn obstacles into opportunities. Today, when things go wrong, I say, "Bring it on!"

I love stories of triumph, success, and self-belief —against the odds, and that is exactly what *Dents in the*

Foreword

Ceiling brings. All of these women are succeeding in their tech careers despite the daily blows, which we now call "*macro*-aggressions," and the major setbacks that have been thrown their way.

Women in tech are following the lead of many great African Americans that have come before us. We have a paved way for us as our foremothers in entertainment, sports, and entrepreneurship have shown us how to overcome. I personally have three remarkable role-models that I'd like for you to remember.

She was denied leading roles and edited out of films when they were screened in Southern theaters. She was told to lie about her race and present herself as "Latin American," because producers believed it would make her more marketable. She could have walked away from her Hollywood dreams. She didn't. She's Lena Horne—Grammy, Tony, and Image Award-winning performer. (May she rest in power.)

She was born in a village in the highlands of Kenya with no formal schooling until age eight. She lost her first university job due to gender and tribal bias, and her first husband divorced her, calling her "too strong-minded for a woman." After referring to the judge who ruled over her divorce case as "incompetent," she landed in jail for "contempt of court." She could have stayed silent. She could have disappeared. She didn't. She's Wangari Maathai, the first African woman to receive the Nobel Peace Prize.

She was raised in poverty and subjected to sexual and emotional abuse. She was fired from one of her first TV news anchor jobs and deemed "unfit for television." She's been criticized for her weight, her hair, and for qualities entirely outside her control. She could have abandoned her work. She didn't. She's Oprah Winfrey. And the rest is history.

Foreword

For the first time in history, we've got a woman of color Vice President, a Black woman on the top of *Forbes* Richest Women in Entertainment list, and sisters in positions of prestige and power at organizations like MSNBC, Walgreens, Microsoft, The Robert Wood Johnson Foundation, and The World Bank. We are living in a pivotal time in history.

With more media attention and opportunity than ever before, it's time to redefine what it means to be a courageous, compassionate, and confident Black woman today.

We've got obstacles to overcome, stereotypes to smash, and trauma to release—no question about it.

I know it won't be easy, but it will be worth it. But my question for you: how big do you want to live? Isn't it time you do something for yourself? For your future? For your soul? For your grandmother, whose prayers that are still covering you?

Quiet the little voice of self-doubt and fear, and let's turn the page together.

xo, Karen Arrington

Karen Arrington

Founder, Miss Black USA / Success Coach / Empowerment Expert / International Philanthropist / Goodwill Ambassador to Sierra Leone and The Gambia

Dents in the Ceiling
Tools Women & Allies Need to Breakthrough

"And at last you'll know with surpassing certainty that only one thing is more frightening than speaking your truth. And that is not speaking."
—Audre Lorde
"The Transformation of Silence into
Language and Action"
from *Sister Outsider: Essays & Speeches*

Introduction

It was the late 1920s, but instead of flapper dressers and dancing the Charleston, my grandmother was growing up in the hills of Appalachia Kentucky. She was number four of eleven children, the daughter of a coal miner and a homemaker—my great-grandparents, who we called Big Daddy and Big Momma. My grandmother is a self-professed "country girl" who loves country music, football, apple pie, oatmeal raisin cookies, root beer, and peppermint candies. I'd call her "spunky" even as she crosses into her 90's refusing to use a hearing aid or walking cane. She still cooks, cleans, and dresses herself, and will even make the occasional appearance in a TikTok dance video with her great-granddaughter. My grandmother grew up riding horses, milking cows, and "beating the butts of those cracker hillbilly White boys that lived across the railroad track." Granny passed on lots of amazing traits such as her famous apples and yams dish, caring for other people, and of course, her resiliency.

"Angel, be strong. You can get through this. You can get through anything," she tells me.

I learned from the best, since this is the same lady who survived the Jim Crow south, the first to desegregate a neighborhood on the outskirts of town in Alliance, Ohio, who raised four Navy men and a PhD daughter on the income of her nurse's and her husband's bricklayer salaries. She stepped in to help raise me after my parents' divorce, and for that, I am eternally grateful. She pulled her strength from a very long line of tough women and passed it on to me. I also picked up the habit that her generation is known for, which is silence. I mastered bottling up feelings, masking my truth, keeping silent, and not voicing my pain. I didn't lose my voice; I chose to mute it.

As a young, Black, introverted, "smart girl" living in a small Midwest city, I had a rather uneventful childhood and not too many friends. I played the "good girl" by most peoples' standards. My family was caught on the fringe of middle-class where we sometimes sat and counted pennies on the living room table to scrape up enough change for me to get a vanilla cone from the ice cream truck, but I also never wondered where my next meal was coming from or if I'd have a new uniform to wear. We always had enough when it came time to pay the $7,000 annual tuition for the private school I attended. We had our financial ups and downs, but it takes a village, and my village was my mom's parents and her four older, Navy-enlisted brothers. They banded together to get me through every piano and ballet lesson, cheerleading practice, after-school and STEM-based summer program, and even the two-week Computer and Electrical Engineering camp in Terre Haute, Indiana.

Introduction

It was a bit lonely, though. As an only child who was never Black enough for my Black friends or White enough for my White ones, I constantly had to explain that yes, I am completely Black, not mixed. (Both my mom and grandmother have a very fair complexion; the assumption was that they must be mixed race.) Private school didn't afford me the opportunity to be "up" on all the latest dance trends, for example. Once, in the front yard playing over my friend's house after school, she and I were two out of three black girls in the entire private school that year, so it was kind of obvious that we'd end up as friends. First grade for me and second grade for her, we were thinking of what game to play next. As a Jheri-curled teen came outside complete with his stone-washed jean shorts, neon t-shirt, and boom-box—it was the 80's after all—the kids in her neighborhood decided on a street-wide dance party.

"What's wrong with you, girl? You don't know how to do 'The Snake'?" said my childhood friend of almost four decades.

"No, what's that?" I replied.

All the kids knew the latest dance moves from watching *Solid Gold, Soul Train,* and each other. Since I had not yet been introduced to these shows, I was the wallflower with no wall to hold up. I got teased mercilessly by the neighborhood kids, which was short-lived once my friend committed to giving me dance lessons after school for the next two weeks. By the next dance party, Angel was out-snaking everybody!

Unfamiliar with current Black culture hallmarks such as dance moves, slang, and never having a Jheri curl always left me feeling a step behind everyone else and on the fringe of both Black and White cultures. However, I am happy to report that by junior high I was successful in talking my mom into permitting me to have a Salt-N-Pepa asymmetrical haircut. I clearly struggled with all the normal identity growing pains and the trying-to-fit-in phases like every other kid. My struggle wasn't unique by any stretch of the imagination, but it was silent.

My first memory of choosing silence instead of being truthful about my feelings came in third grade. It was just after a Black History Month lesson in Social Studies. The 3:15 bell rang, and we all filed out to the school bus and car pick-up line. Granny was always third in the ride pick-up line in her pale mustard yellow hooptie, waiting with a big smile and a funny story about the patients she cared for as a nurse. On our way out the building, Zack, the most athletic, tall, blond-haired kid in class, turned to me unprovoked and said, "You know all Black people are niggers, right?" Well, that's what my grandpa says; if you're Black, then you're a nigger. You know you're a nigger, too, right?"

I recall his best friend Gary, the light-skin Black version of Zack, turning around as both our mouths were agape, our eyes wide in shock. That moment when our eyes connected felt like a floodgate had opened and all the stories and feelings that we, as two of the four Black kids in class, had just heard of slaves and slave masters came rushing in, filling the space between Gary, Zack, and me.

Gary broke through the time trap by shoving Zack hard and saying, "Shut up, moron. Why'd you say that?"

Zack replied, "Well, she is!"

Me... just me? I'm a nigger? What about Gary? Are only Black girls niggers? Is that how Zack interpreted his grandfather's racists views? Why did he say that to only me? Why did he say that now? What did I do to elicit his seemingly nonchalant-lined-with-a-sliver-of-rage comment? Zack and I had never interacted, so why was I a target? All this and more

Introduction

swirled in my eight-year-old brain as I opened the creaking passenger-side door and climbing in to face my pie-faced grinning Granny as she threw the car in drive and happily inquired, "How was your day?"

That was my moment of truth. My lips parted, but my tongue remained still. I refrained from launching into the story. Instead, I sat confused, angry, hurt, embarrassed, and near tears until we got home. That moment, like so many countless others, is how I choose to respond to pain well into adulthood. Later, facing unconsciously biased coworkers once I arrived in my tech career, I stuffed down the hurt, confusion, and embarrassment of being treated unprofessionally (like being yelled at like a child in team meetings), endured the hurt of the sexists jokes, and buried the pain that accompanied having a racist and sexist supervisor.

"Silent" is exactly how I would label much of my life. Not bringing sound to the painful moments, instead concealing my ideas, passions, thoughts, desires, wants, and needs. But why? Perhaps because despite the feminist movement of the 60s and 70s, it didn't quite reach folks like me. "I am woman, hear me roar," and whatever rhetoric Gloria Steinem spouted didn't apply to me. I didn't even really know who Angela Davis was until I met her when she came to speak at a Women's Council Meeting at the University of Pittsburgh my junior year. I decided to minor in Psychology and Women's Studies to round out my Information Science degree. This signaled the point of my enlightened period. I immersed myself in learning about the feminist movement; I took a few classes in Black Studies and read women slave narratives by Harriet Tubman, along with Angela Davis, Toni Morrison, and Zora Neale Hurston. Their stories made sense to me. I had more in common with women from two to three generations before my own than I did with the White women just eight to ten years my senior. These Black women's stories sparked something in me that has taken two decades to finally come to light. I was in pain, and my pain matters. Yes, after 20 years in corporate America, I'm learning what it means to be authentic and share my story.

The goal? To create space for all women by illuminating the path toward inclusion. I want my companions to know they are part of a tribe where they can be their smart, powerful, funny, quirky, weird, incredible, passionate, assertive, confident, sassy, get-shit-done selves. They will not have to worry about contouring to a stereotype or fighting against it because we welcome you, we love you, we need you, we bless you, and we encourage you! No longer will I allow my voice to be silenced out of fear. I will speak my truth because it will not only set me free but will free others as well. What is this freedom we need, you ask? Harriet Tubman led hundreds of captive slaves out of physical bondage, but the emotional and mental journey to freedom continues for us to this very day.

The emotional weight that African American women carry shows up in the classroom and the boardroom and every stop in between. The overlooking, ignoring, and excusing of disrespectful and unprofessional behavior is part of our everyday experience. The weight of the silence and burying of feelings can get so heavy that the psychological and emotional impacts can lead to negative physical outcomes. But we are still here and still making dents in our concrete ceilings. It is both a testament to our resolve and a miracle.

As a woman of color in the male-dominated IT world, my previous work experiences had clearly been filled with microaggressions as if they were the norm, such as limited support from my senior leadership and ignored accomplishments. I walked on eggshells and worked to conform to the IT industry, complete with my standard-issue khakis and collared polo-shirt. I was saddled with feedback that I was "too passionate," "not technical enough," and "always trying to move too fast," or, for some executives, "not moving fast enough."

Now that I'm with a new company I have one major problem. What do I do with all the emotional baggage that I've collected over the years? 20 years ago, Erykah Badu sang me a warning about carrying my bags around. The only issue was that I didn't know where to put them down. I didn't know how to start the process of releasing the weight. Then I remembered the leadership program my mentor had invited me to and thought maybe, just maybe, an all-African American female leadership program filled with women who were mid-career in their IT roles could help. I quickly negotiated the sponsorship of that program into my on-boarding package with the new company, and just like that, I was in!

Enter EMERGE.

The Information Technology Senior Management Forum (ITSMF) leadership EMERGE Academy's mission is to hear her voice. The program's goal is to increase the representation of women of color at the senior level in the technology field through a series of professional workshops, supportive environments, and model leadership with an eye toward the unique experiences women leaders of color have in the workplace. It is with thanks to EMERGE Director Kailei Carr that this book was born. It was Kailei with whom, after an EMERGE session late one evening, I shared my pent-up feelings about previous corporate hurts. She introduced me to the term "emotional tax." Emotional tax is described by Drs. Travis and Thorpe-Moscon of Catalyst as when "you devote time and energy consciously preparing to face each day, which you know comes with the potential for

Dents in the Ceiling 18

Introduction

large and small acts of bias, exclusion, or discrimination. This requires daily, not occasional, vigilance." Kailei stressed that emotional tax is real and most likely what I was suffering from. The cure, she said, was to "journal."

I scoffed, "Oh, Kailei, I don't write. I could barely keep a diary in junior high." *Plus, when would I have time?* I asked myself.

I have a husband, two kids, a full-time job, was an adjunct instructor, plus, an active sorority member, an ad-hoc caregiver to my 90-year-old grandma, all the while chairing two committees for two separate conferences at the time.

"Um, no, Kailei. No time to write."

That is, until I experienced another incident of bullying. It was relatively minor in comparison to my previous experiences. However, it triggered in me the old hurts and old times that I'd ignored, turned the other cheek, and practiced my best Michelle Obama—you know, going high when they went low. Well, I was so high that I was practically flying at times.

That night I logged off my laptop and opened my notebook. I managed to pour it all out. I wrote down every turned-cheek and blind-eye moment from when I was a young (fresh out of high school) intern at my first IT company up to that very moment in my IT career. My pen couldn't get the words down fast enough as I retraced my journey through the halls and conference rooms of corporate America. I sat back, shocked at how much I'd forgotten yet finally giving words to my pain.

As I sat in my basement in the early hours of the morning, I pictured Mark, Jeff, and Sundar (2020 top Silicon Valley CEOs) scratching their heads, wondering why their diversity initiatives and training investments were bleeding money. Well, not to pat myself on the back too much, but I do have the answer! I'll be kind and give you the conclusion now—it's not just me! It was time that I, along with other African American women, broke our silence. 10 months of collecting 30 personal narratives from African American women who work in STEM, all in corporate America (with one working in a research lab) all reveal the same themes I'd found that night in my basement on the pages of my feel-good notebook and more. So much more!

Don't believe me about how challenging corporate America is for African American women? Can't phantom a "don't touch my hair" sign outside someone's cube wall? Imagine being hit by a phone, kissed on the mouth by your brand-new boss on the first day of work, or a supervisor asking you how much sex you're having. Those stories and more are what you will find in *Dents in the Ceiling*. And don't worry, I changed the names to protect the innocent... and the guilty.

> "People resist by . . .telling their story."
> —bell hooks

A Thematic Analysis Combined with a Narrative Analysis Approach

Where does one start when they made their bones working in corporate America for two decades and not as a full-time researcher? Why, her network, of course, along with a little intervention from the good Lord above! A couple of weeks into setting up interviews to meet with African American women about their experiences in IT, my coworker mentioned that his daughter works in a lab at the prestigious Spelman University, and she wanted to interview me for her research. What on earth was her research about to where I would be a subject? Why, African American women working in STEM fields in corporate America, of course! I told you, the Lord works in mysterious ways.

Long story short, I had the privilege of meeting the lab principle who helped me strategize on my work and research. She explained that there are many approaches to qualitative research analysis, studying experiences and meaning (versus quantitative research, or numbers and statistics). So, hold on a moment while we delve into the world of qualitative research analysis while I explain my research approach. The lab principle introduced me to the Grounded Theory Approach, Narrative Analysis, along with other approaches to qualitative research and coding methods.

I began with a thematic approach, a set of themes I saw in my personal journaling, then followed up with the narrative analysis approach to allow for new themes to emerge. I started with themes I had in mind already based on preliminary research and personal experience. I was seeking women who worked in technology in corporate America to see if their experiences were different or like my own. In general, as researchers review the data they collect, repeated ideas or concepts become apparent. These ideas/concepts are said to "emerge" from the data. The researchers tag those ideas/concepts with codes that succinctly summarize the ideas/concepts. As more data is collected and re-reviewed, codes can be grouped into higher-level concepts, and then into categories. These categories may become the basis of a hypothesis or a new theory. This method of analysis is a slight departure from the traditional scientific model of research that we were all taught in school, where the researcher chooses an existing theoretical framework, develops a hypothesis derived from that framework, and only then collects data for the purpose of assessing the validity of the hypothesis; in this case, I had preexisting themes to study my hypothesis.

Notably, thematic and narrative analysis approaches can be combined. Most important to me, according to Melissa Petrakis, research practitioner mentioned in her summary of the various analysis approaches, the focus of the process is how one looks for themes.[1] If you start with preexisting themes, you are on a deductive path; however, analyzing the narrations is an inductive approach. Markus Jannsen states, "I think you could use a deductive-oriented thematic analysis to search your data for cases where the themes you are interested in are represented and then analyze these cases with narrative analysis."[2] Based on the discussion between Melissa Petrakis and Christoph Stamann, I combined both approaches, which meant I used my themes as a hypothesis but looked for additional themes and experiences to emerge from the data.

My hypothesis: there are more roadblocks and obstacles that occur for African America women in IT between the journey from internship to the C-Suite that corporate America isn't willing to acknowledge; if we have any hopes of mitigating them to make significant

---

[1] Melissa Petrakis, "Re: What is the difference between narrative analysis and thematic analysis? Is thematic analysis an approach of narrative analysis?," *ResearchGate*, 2017, researchgate.net/post/What-is-the-difference-between-narrative-analysis-and-thematic-analysis-Is-thematic-analysis-an-approach-of-narrative-analysis/58ca7c3ef7b67ef27153937f/citation/download.

[2] Markus Janssen, et. al., "Qualitative Content Analysis—and Beyond?," *Forum: Qualitative Social Research*, vol. 18, no. 2, 2017, doi.org/10.17169/fqs-18.2.2812.

advancement, we must know what the issues are in order to propel our melanin-positive sisters forward. So, let's go find out where these impediments are, publicize them unapologetically, and address them to increase our numbers of leading Fortune 500 tech companies! For starters, let's just get on the map!

Through virtual calls, home-schooling, my husband's tragic ATV accident that caused a severe concussion, multiple family deaths, and managing a full-time job while adjunct teaching, I interviewed women at lunch, in the late evening, and on the weekends. I asked for specific incidents of discrimination such as bullying and sexual harassment, but I got much more! So many incidents of microaggressions were shared with me that I dedicated a whole chapter, which I've called Macro Hurts, since the totality of scars is anything but micro.

The interviews were transcribed and then uploaded to a qualitative analysis tool. The software permitted me the ability to read the interview transcripts all in one module, highlight key words that repeated from the ladies' responses (or coding), then perform the data analysis as described by Petrakis and Stamann above. Dedoose allowed me to visually "see" the intersections of being "the only" to bullying or being treated like the maid; when there was a high existence of co-occurrence of two major themes, the chart turned red to visually draw my attention to critical connections. The themes that emerged from these interviews, such as being the maid, bullying, and sexual harassment, absolutely exist. However, as researchers of Narrative Analysis predicted, new themes also emerged. For example, fear, White female supervisors/coworkers being inadequate allies, being forced to move jobs, resiliency, working for a "Greater Purpose," and allyship were prominent themes throughout the interviews.

Introduction

How to Read this Book

Read this book like an explorer on an expedition. If you are a minority, female, ethnically or racially diverse, or fall into an underrepresented class, then you are the primary explorer. I will lead you through a deeper understanding of what it is like being the lonely only, being bullied, ignored, treated like the maid on your team, sexually harrassed, and of course what it means to deal with a mountain of microaggression. Upon concluding your journey, you will follow the clues I lay out for you to fill your toolbox to the brim in preparation for a life marked by strength and resiliency. If you are not in an underrepresented class, I invite you to explore with us and experience the journey for you to become an active ally.

The statistics and data around how Black women are treated in corporate America can be difficult when you read with your heart and not your head. We know that African Americans are underrepresented in IT and STEM-related fields, often because of perception, not performance. While metrics around minority representation in IT are slowly changing for the better, recognition of African American women leaders continues to lag—woefully! In fact, we are losing ground. African American women hold 1.6% of Vice President roles and 1.4% of C-Suite executive positions. The first African American female CEO was Ursula Burns, serving as Xerox CEO for seven years; two years later we got Mary Winton as interim CEO for Bed Bath & Beyond but only for less than a year, when she was replaced by a White man. Two years later, it appears we will get two more CEOs and possibly counting? I don't know about you, but I need this rate of change to be accelerated...now!

As I sit here writing, today this country just elected our first woman of color Vice President. I am filled to the brim with hope that change is finally here. Despite the enthusiasm, make no mistake that there is a great deal of work ahead! The presidential election resulted within a very uncomfortable margin. As America continues to crawl closer toward equality, many women in management positions still struggle to take the next step into upper senior executive and C-level roles.

Knowing these statistics is one thing, but allow yourself to feel the sentiment; feel the emotions we too often hide, push way down deep, or ignore. After you feel the sadness, disappointment, pain, do not stay there. You are not stuck or left without options; please do not pick up the victim card in this journey. Use the clues to move to action and think introspectively about how you show up and how your own self-criticism and past hurts, as well as how negative thoughts play a part in your work experience. Remember, you are in a working relationship (key word "relationship") with your peers; you are not perfect, and neither are they. Own your developmental areas in addition to acknowledging the bias and stereotypes that others have about you. It is extremely important to your personal and professional advancement. The viewpoint in which you read this book is by using the "and" view, so think to yourself: there are unconscious biases and discriminatory processes inherit in my work environment *and* I have some development areas in which I need to receive training, coaching, mentoring, or counsel in order to grow and show up as my best self to overcome them.

Introduction

Dents in the Ceiling offers more than just validating experiences that African American women have in corporate America. It is also very important to know that we are not by ourselves on this journey. Others who do not fall into an underrepresented class in American society are following along to learn more about what African American women specifically endure in the workplace in hopes of obtaining empathy and becoming active allies. At the end of each juncture, clues will be provided that are practical recommendations for addressing the issue at hand for both audiences. Clues help minority women survive and thrive given their work circumstances. Whereas clues for any other population serve as clear action steps that can be utilized to offer support.

Our final chapters end with assistance from the pros—counsel from experts in the fields of executive coaching, health, and wellness, and those who have researched and worked tirelessly in the field of advancing women at work. We get sage advice on career development along with self-care to maintain our health and well-being for the long road ahead. Many of us have jumped off the company ladder early to start our own businesses or have left our STEM careers completely. So why stay? Listen directly to the interview participants about what they have set as their North Star, the North Star being what keeps them in tech working for their respective companies. The road is rough, stony, and black at times, which can render you numb to your pain and without a voice, but by reading this book, you'll collect a variety of resilient practices to help you along your career journey to speak power to your truth.

Prologue

I had just introduced myself, provided an overview of the agenda, and was engaging in idle chatter as the 9 a.m. meeting began. My seven-month pregnant belly was just able to fit somewhat comfortably at the long conference room table, and 25 business unit heads and a handful of IT leaders were in the room. It was a huge room that could seat about 70-80 attendees. I was hosting two vendors and four different business units in a two-day vision-setting and requirements-gathering planning session for a new system upgrade, one of the largest in the company's 30-year history. Of course, I was the only African American, as well as the only ethnic minority in the room completely. I sat up tall with a plastic smile on my face as I adjusted, kicking off my pumps due to my ever-swelling ankles. As the Program Manager, I began our first round of discussions. No sooner than 15 minutes into the session, a coworker, John, came racing in with a panic-stricken look on his face, running up to me to whisper in my ear that I'm needed in Conference Room A immediately. I politely excused myself and asked my colleague to my right to take over meeting facilitation while I tended to an emergency. On my way out the door, I noticed that no one had blinked an eye at the mysterious way John had entered the room and that my exit was completely ignored. I wondered if getting there at 6:30 a.m. to prepare had been worth it.

I walked briskly behind John to Conference A and stood in the doorway of a jam-packed room of about eight team members from the other project I managed, all looking at the floor or looking at me with wide, fearful eyes. Before I could utter a word, I was met with the lead Systems Manager yelling at me at the top of his lungs.

"Angel, you get in here right now and run this damn thing! This is YOUR project, and you need to be here to tell these people what to do!"

The Systems Manager was literally red from the neck up, which was a stark contrast to the pale white ghosts around the room. It felt like walking up to a crime scene and being the one accused of causing the dead body outlined in chalk on the floor.

I stammered, "What is going wrong?"

My mind was going a thousand miles a second, so forming complete thoughts and quickly "taking control" of the situation was out of the question at that moment.

He demanded to know why I hadn't been there, since it was "my project."

He yelled, "No one knows what to do, and no one started the fucking testing scenario on time."

A few more curse words and another "this is your project," I finally gathered my wits. I plugged into my emergency crisis training from my brief stint as an

American Red Cross fire volunteer. Speaking calmly, as if to a child having a temper tantrum, it was clear that the Systems Manager had forgotten about the competing, very large, very visible, very company-wide, high impact project that was being planned at the same time as the second round of mock-trial testing on our financial system upgrade. I reminded him that we'd determined two weeks ago that our company was not going to invest in a cloning machine for me to be in two places at once. He was supposed to lead the mock testing while I led the big-ass, revenue-generating, two-vendor coordinating, all-executives-from-Supply-Chain-to-HR meeting down the hall. His response? "Well, I don't care, this is YOUR project, and you need to be here and tell the testers what to do!" I asked him to excuse himself so he and I could talk privately and resolve this, but he responded with, "You don't want me to come out there. I don't know what I will do!"

Bam! I had this insensitive, racist, misogynist jerk right between the eyes now! He'd just openly threatened me in front of a room full of people. I turned on my ever-swelling heels and stomped down to my supervisor and his supervisor's office, quickly telling them what happened, and that Jerkface wasn't willing to resolve the issue with me directly. I let them handle it and went back to my meeting. Inhale, exhale... deep breath. Plastered smile neatly in place, I walked in the room, resumed facilitating for the day as if nothing had happened. In between breaks, I documented exactly what had been said in preparation for a formal HR complaint.

All the microaggressions, comments, and jokes were going to finally catch this guy because I now had something tangible—with witnesses! I went home that day with a healthy mixture of anger, embarrassment, and glee. I hit "send" on my formal complaint at 6 p.m.

sharp so it would be in the inbox for the Head of HR in the morning.

You know that feeling you get in the pit of your stomach when you hear the tap-tap of the first drops of rain on the porch the morning of your huge outdoor birthday party with 20 of your closest friends? That deep, sinking, gut-punching blow of all the air being forced from your lungs when the full weight of disappointment sets in? That's what was happening to my seven-and-a-half-month pregnant belly when I sat across from HR, my supervisor, and the ignorant, disrespectful, White male Systems Manager, as the Head of HR explained that his punishment would be mandatory HR training to be completed by months' end. (She never stated what type of training—anger management, maybe. And I didn't ask.) The Head of HR also mandated him to give me a verbal apology for embarrassing me.

He apologized, I suppose, but I didn't hear it. I was too busy thinking about whether JCPenney carried maternity business suits now that interviewing at seven-and-a-half months pregnant was about to commence.

Chapter 1

Visibly Invisible

"Recognize and embrace your uniqueness. I don't think the ratios are going to change anytime soon. But, I don't think it has to be a disadvantage. Being a Black woman, being a woman in general, on a team of all men, means that you are going to have a unique voice. It's important to embrace that."

—Erin Teague
Director of Product at Yahoo

Welcome to the Valley of the Lonely Only. One would think that being the only melanin-positive person in the room would make me stand out like a sore thumb. Yet somehow, I can still be rendered practically invisible. The biggest issue for me is the mental hoops that I play with myself as I sit there and listen to the political tennis match taking place between team members all while a different internal banter is playing. Trying to decide if I should jump in or not. If I should keep my mouth shut and stay out of it. Then I render myself invisible with no voice, no opinion, fearful of leaving the impression to others around the table that I have nothing to add, nothing to contribute. On the other side, if I jump in to offer a comment, question, or opinion that is off-base, then I could become irrelevant, dim-witted, or just plain stupid. This is the dichotomy of physically standing out yet being hidden at the same time.

54 percent of Black women say they are often the "onlys," meaning that they are the only Black female or one of the only Black people at work. Black women who are "onlys" have an especially difficult experience. We are very aware of the fact that we may be representatives of our race, and we are more likely than the "onlys" of other racial and ethnic groups to feel as though our individual successes and failures will reflect positively or negatively on people like them. Jonathan Capehart commented on this phenomenon in his podcast interview with Rosalind Brewer, COO of Starbucks at the time, in December 2019: "I asked Brewer to talk more about what it's like being 'the only one,' the only African American in rarified and some not-so-rarified spaces. She said, 'It absolutely does not feel good. And I will tell you that it's actually quite lonely.'"

In their report *The State of Black Women in Corporate America* the organization Lean In found that being "the only" leads to a sense that we are constantly under scrutiny. Black women who are "onlys" often report feeling closely watched, on guard, and under increased pressure to perform.[3] In an accompanying graphic of the information gathered from survey respondents, it is evident that Black and White women have a very different experience in the workplace. This

---

[3] *The State of Black Women in Corporate America*, Lean In, 2020, Lean In.org/research/state-of-black-women-in-corporate-america/introduction.

report also illustrates why it is important not to lump all women into one category. In fact, at my company, when we realized that the additional characteristics of race and ethnicity were playing out so strongly, we had to create two separate Employee Resource Groups—one for Women in IT and one for Women of Color. From the outside looking in, you would question the need for two separate women's groups, but studies like the one conducted by Lean In along with others clearly show that race is a stronger influence on women than gender.

A fellow African American working in STEM in corporate America, Monica Brown, knows the valley-experiences of being an "only" so well that she dedicated an entire book to it. Being "the only" carries a heavy weight and is a burden for us, really. This emotional and mental "extra" is called emotional tax. We'll explore this "taxing" situation more in Chapter 6, but for now, just know that we are stressed while trying to prove our worth. This occurs regardless of how many degrees or certifications we have. We not only feel as if we have to work twice as hard to get half as far, but we are also explicitly reminded by our parents, grandparents, peers, and community.

The National Science Foundation reported that in 2016 alone, Black women earned more than 33,000 bachelor's degrees in science and engineering, and 24% of doctorates awarded to Black women were in STEM. However, the same report showed that in 2017, only 5% of managerial jobs in STEM were held by Black women and men combined. In 2020, Catalyst.org found that White men still hold 62% of managerial positions, while 32.5% went to White women, and 3.8% were held by Black women. We don't have to cite these statistics; we know them because we live them every day. We are easy to spot in a small conference room, and we are easy to spot at every company kickoff, product end user conference, and technical trade show. A quick scan of the room allows me to mentally calculate the numbers. For instance, if a conference had reported 1,500 attendees, and in the great hall during day two of the keynote speech, I could skim the crowd and count on two hands and one foot how many of "us" are in attendance (being generous to include those who look ethnically ambiguous), then I feel like an "only."

The challenges still abound once we are in the door. Many of us experience the "pet to threat" syndrome. It's cool being the new kid on the block. Coworkers and others in the organization tend to put their best foot forward and are consciously kind, polite, and helpful when you first arrive. I liken it to dating.

When starting a new relationship, you enact your representative. You watch how you chew your food at dinner, mindful of letting the other person talk about themselves, listening intently to the childhood stories or antics of their best friend's love life. You volunteer to help them move to another apartment or cook a meal, and of course you let them pick the movie—most men do all this and more. We do our fair share to keep our make-up intact and our hair styled, and we wear our good underwear, just in case. Slowly after the relationship matures, our representatives come out less and less. The first argument is a great tell-tale sign of who the person really is when they can't get their way. This can be easily translated to the work world.

The first team meeting is usually filled with nervous laughter when the supervisor says, "Everyone be on their best behavior; we don't want to scare {insert new person's name here} away." It's funny only because it is true. Over time, the newness starts to wear off and by the end of the first year, you notice that your ideas for process improvement aren't met with as much enthusiasm as previously, or the amount of red tape to get approval for a new person on your team grows and you wonder what is happening. Perhaps no argument occurred, but you are now starting to get comfortable in your new position and have an increased feeling of "I got this." You have more ideas for changes in the organization and are ready to take on new challenging work assignments, but the lead balloon of reality drops on you like the house did on the Wicked Witch in *The Wizard of Oz*. Depending on how severe the argument or how hard you push for change, you notice the chilly reception in team meetings, the constant cancelling or rescheduling of one-on-one meetings, and people that had been your go-to helpers avoiding you and not returning your emails or phone calls. This is the point in which the analogy of a personal relationship to a professional one diverges. In a personal relationship, flowers, cards, candy, or perhaps a midnight walk to the park to talk things out can easily shift the tension and anger from the disagreement and lead to reconciliation. Except the African American female who has lost her shine in the workplace can't send cards and candy. Unfortunately, this is the point in which we move from pet to threat.

University of Georgia Associate Dean Kecia M. Thomas coined the term "pet to threat" as an emerging trend after conducting year-long research of over 35 African American professional women. One author, Erika Stallings, follows Thomas' work and captures a story of one woman's transition from pet to threat at her law firm.

> "After another 12 months of being underutilized, I decided to look for a new opportunity to get the litigation opportunities I had been missing out on. The process of giving notice and the conversations that unfolded during

my final two weeks were miserable. There was a general attitude from my employer that I was ungrateful and wrong to complain about my lack of advancement. I didn't know it at the time, but I had likely been the victim of a workplace phenomenon known as "pet to threat." This happens when women, typically Black women, are embraced and groomed by organizations until they start demonstrating high levels of confidence and excel in their role, a transition that may be perceived as threatening by employers."[4]

A healthy mixture of stereotypes, unconscious bias, and the sentiment of "just be happy to be here, shut up, and do what we hired you to do" leaves little recourse for the future career prospects for a woman. Why a threat? When we challenge the status quo, seek to change processes that the organization has "always done," or push for a promotion, we become challenging to those who have been there, which causes discomfort and feelings of being threatened. This situation has a few courses of action: stay and endure, leave, or create a third option for yourself which includes work outside your current organization to show your value and increase your sense of worth.

Unconscious bias (a new-age, fancy term for stereotypes), the "comfortability" factor (we just read what happens when people become uncomfortable and feel threatened), and informal networking are what I have personally found to be the big three that make our workplace experience especially challenging.

---

[4] Erika Stallings, "When Black Women Go From Office Pet to Office Threat," ZORA, Jan. 16, 2020, zora.medium.com/when-black-women-go-from-office-pet-to-office-threat-83bde710332e.

Unconscious bias means that our human brains are lazy and love to categorize things, people, and anything we encounter in our environment as friend or foe, then sub-categorize from there to make it quick and easy for us to survive. To make my day as efficient as possible, I make hurried decisions. I have to think fast and react quickly. The speed of technology often outpaces everyone, so IT leaders must categorize and go!

Ingrained stereotypes help us move fast, but this also applies to people as well. We categorize people based on how they look, speak, and behave, and these assignments within specific categories are made in seconds—no real thought necessary. For example, if I have a box in my mind for "White men," then whenever I encounter a White man, he goes in the "White man" category in my head. The more I interact with him, the more granular I can get in terms of "ally White man" or "racist White man." The same categorization occurs for Black women. Some negative boxes can include "mammy Black women" or "jezebel, loud, ghetto, uneducated Black women." My sisters and I have been working extra hard the last several decades trying to create alternative boxes, such as "educated, intelligent Black woman" or "leader Black woman."

The comfortability factor can also be called the "likability factor." Being in the likability zone is ideal since this is where the magic happens. This is the zone where unwritten rules or norms of the organization are shared, where you find out the politics of who are friends with each other, who you should never challenge openly in a meeting. In essence, having the "likability factor" is when the person gets to know you on a more personal level and can trust recommending you to more senior positions. For example, it is just easier that Sam the Senior Manager will remind John the CIO of himself at a point in his career, enabling John to like Sam and want to help him through mentoring or sponsoring. Unfortunately, there are quite a few degrees of separation between John and Tameka that require a great deal of effort for both to navigate in order for Tameka to get into John's "likability zone." It isn't impossible; it just means one relationship is effortless while the other one requires intentionality.

Sometimes, being liked is all you need, especially if you are a White male. I have seen many charismatic, good-looking, well-polished White men climb the ladder —I mean, get pulled up the ladder—only to fizzle out at the top because they lacked the experience they needed to learn along the way. That is certainly not true for all of them! Not everyone is born with a silver spoon that gives them a "get out of jail free card." I've been equally saddened when certain White managers, directors, and VPs did not get the well-earned step up I thought they deserved. My point is that in the corporate environment, the number of hoops for them to jump through are significantly less than for an African American woman.

Qualities such as competence in your job, building your team up, being a good leader, uplifting your peers and staff, plus excelling at being a proactive, innovative, and technically sound leader are just icing on the cake for White men in the corporate world, but all that and more is considered table stakes for us. As a double minority, double expectations exist for us to even have our name said in the room during performance evaluations, top-talent reviews, or considerations for an extra bonus (which is a thing, by the way) or stretch opportunities. For our counterparts, their potential can land them the job versus our performance landing us ours.

Last but certainly not least, informal networking is a major factor in ladder climbing. Opportunities to socialize at the bar after work, on the golf course

Saturday morning, or during a Sunday afternoon football game are locations where coworkers can talk outside of work and get to know each other personally. Lots of people code-switch at work, even able-bodied, heterosexual White men, but letting their "hair down" doesn't cause them as much anxiety as it would for me or any other minority. Informal networking provides an amazing chance for a person-to-person connection to form which gets you in the "likability zone." According to the Neuroscience Leadership Institute, informal networking will most often land the White male in the Similarity Bias bucket. Similarity Bias is the preference of those like us over those who are different; an in-group/out-group mentality. Similarity Bias essentially accounts for the "likability factor" because there are fewer degrees of separation for them than there are for us.[5]

    I know I shouldn't care. I should just speak my mind and let the chips fall where they may, right? I'm constantly watching for just the right time to jump into the double-dutch conversation occurring during a team meeting. The times in which I'm silent are because of the debate going on in my head as I sit between the angel and devil on either shoulder. The angel telling me to speak up, share my often-diverse and divergent opinion, while the devil is reminding me of the need to watch what I say, to not always be the dissenting voice, and to avoid coming across too forceful or meek. Every time I speak, I must proceed with caution. Then I do it. I work up the courage to voice an opinion, a decent one, I think to myself. One that offers a different perspective but not too far left.

    What happens?

---

[5] Chris Weller, "The 5 Biggest Biases That Affect Decision-Making," *Your Brain at Work*, April 9, 2019, neuroleadership.com/your-brain-at-work/seeds-model-biases-affect-decision-making/.

Nothing. Absolutely nothing.

Well . . . not in the meeting, anyway. Rarely does anything happen during the meeting. It's usually after the meeting that I'll get a call or text message to the effect of, "Good idea, I agree with the point you made. Not sure why we didn't go with it?" Or my favorite, "I was with you. You were right." I'm left thinking, Okay, so I'm not stupid. I can add value, but siding with me publicly isn't in the cards, huh?

My recommendation is to acknowledge then encourage. For example, about five to six years into my career, I landed my first female supervisor. She's a great leader, sharp, fair, and most importantly, she supported me. When I made a recommendation, she backed me up 100%. We found a connection since we both had attended the same college in undergrad and could share fond memories of the local restaurants, hang-out spots, and culture. She was a White lady who fought her way to becoming a manager in the company, in which 80% of managers were White males. She was one I considered to have "earned her stripes," and after reporting to her for over a year, I had made it into the "likability zone."

As would occur often, she approached me with an opportunity. All her opportunities up to that point had panned out to be good ones for my career, so when she expressed her intention to increase my exposure to our department's senior leaders, I was all in. I was staring down the face of an approaching promotion cycle after failing the first one. Why? Not enough exposure. Senior leaders knew I was a solid project manager, but they needed to "see" for themselves before signing off on my promotion to Senior Project Manager. So, turning into my advocate, she said she would go to bat for me during the next promotion cycle and therefore needed solid ammo. This meant I needed to be seen. I needed to be in meetings speaking up, sharing ideas, making proposals, and volunteering for assignments. All the usual stuff I did behind the scenes, only this time, I needed to be "in front of the camera" doing it.

Thanks to my manager, I was invited to the year-end departmental business planning meeting. Typically, I would spend several late nights and long weekends every September number-crunching and performing the data analysis to prepare my supervisor and director for the meeting, but this time, I was in it! All the department managers were there with their inch-thick stack of projects and project data, armed to the tooth to defend their proposals. I didn't have much. I had two projects, but they were sizable. Although I was a bit unsure on my staffing calculations, I was confident that at least one of my projects would make the master list.

The Senior Director, Savannah, led the meeting. An average height, average built, blond-haired White lady who I thought was too young to have such a senior position, but I didn't know her story, so maybe I was just jealous; she facilitated our department project list creation session. I did take note that she always, and I mean always, wore high heels. The click-clack of her heels was memorable, as it could be heard even on the thin carpet of our basement floor cube area at least once a day—usually for her coffee run to our floor about mid-afternoon.

I caught on to the rules of the meeting quickly. Savannah called your name to speak, you shared your list of top recommended projects to be put in the budget for the next year, everyone quietly listened, asked a few clarifying questions, and you concluded. It was time for the next person to speak after Savannah thanked the previous person for their contribution, and she nodded to the next person to share their project proposals. This session was much more formal than I'd ever imagined, but I still looked forward to my turn.

It was round-robin style, so you spoke in clockwise position around the table. Savannah started the meeting then nodded for Ken to proceed with his list. After Ken spoke, Savannah thanked Ken, then Lance provided his project recommendations, and Savannah thanked him, then Kimberly debated Lance's projects. After the heated discourse, Savannah thanked Lance and Kimberly. I piped in with my project considerations, and you guessed it, silence. A nonexistent "thank you." James popped in to ask about the timeline of the budget approval process, Savannah answered James, and the meeting concluded. Not as climatic as you were hoping for, huh? I completely felt the same way! As we walked out, Lance stopped me outside my cube and said, "I wonder what they will do with your two projects. I need your ADS project to finish before mine." This was the point in which the rapid blinking and pursed lips began and continued throughout the remainder of the day.

By the end of the day, I felt totally defeated. Did my entrance into the land of exposure hurt my chances of promotion? Was Savannah underwhelmed by my stats? Was my argument not compelling enough? Then on the social side, I wondered, why didn't she at least thank me like she had thanked all the other presenters? Wasn't it odd that she never noted my new presence in the room? Had others been wondering why I was there to present these two projects when my supervisor easily could have represented our whole team's work? These are the types of questions I gnawed over endlessly.

I was too scared to ask my supervisor how I'd done, so I sat and waited two whole days until our Friday mid-morning one-on-one session and braced myself for the feedback. The response was, "Not bad." I had solid arguments, and everyone knew my ADS project needed to make the cut, so I had no need to worry. Without any prompting, my supervisor shared that she told Savannah she was inviting me to the meeting that was usually reserved for managers to give me exposure and help me learn "how the sausage is made" when it came to the final budget. Since I'd worked tirelessly and without complaint for three budget cycles on the numbers, it would be fair to include me in on the next step. She said Savannah had nodded and walked away. Without a good mentor in my corner to counsel me during this time in my career, I decided to take the bull by the horns and make small talk with Ms. High-Heels the next time she made her mid-afternoon trip to our work area. Monday at 2:45 p.m., like clockwork, she click-clacked her way past my cubicle. I followed her to get water and chat about the budget meeting. To no avail, she nodded, said the final budget list would be out by Wednesday and that I could get it from my supervisor. As she click-clacked away, I mentally started running down a list of other Directors and Senior Directors in our area with whom I needed to get into the "likability zone" before promotion time, and I mentally drew a line through Savannah's name.

My situation is not unique, as you will read in the following excerpts from respondents of my interviews of African Americans who work in STEM in corporate America. These women shared their personal experiences, and being "the only" was a prominent theme.

Frontline Tales

> I was the only Black female in that organization. So, I was given a stigmatism of, "She's difficult to work with."
>
> —Laura, Senior Program Manager at a Multinational Engineering Company

> I said, 99.999% of the time, I'm the only woman. I'm the only Black woman. And I've spent a lot of my career making White men comfortable." I said, "I'm not in that business anymore. So, if they're not ready for me to show up fully as me, this is not going to work."
>
> —Alicia, Freelance Agile Consultant

Being "the only" can be quite disheartening, leaving us with feelings of loneliness and hopelessness when we don't see anyone that looks like us in executive roles or even our own peer group. It can also stunt our career growth. Negative feelings left unchecked can lead to hostility and falling into the "angry Black woman" trap.

> There is no support. You look around and you—well, let me say I look around—and I don't see people that look like me in positions that can lift me up. There are very few Black people at the top.
>
> —Alicia, Senior Agile Coach and Trainer at a Mid-Size Consulting Agency

"Oh, you need this many more years. You're not ready yet. You need this much more experience." And I do attribute that, in ways, to it being a male-dominated industry. I also attribute it, in ways, to the fact that I am African American, and being in that company, being the fact that the culture there was very apparent who was in the leadership ranks, very slow to change, who would grow to become a leader in that company.

> —Diane, Business Analyst at a Fortune 500 Pharmaceutical Company

Cybersecurity is primarily composed of older White males. And so, when I came in right out of undergrad, I got a lot of hostility. And being young and naïve, I really didn't know where it was coming from because I've always been the type of person, "Hey, I just want to learn. I'm here to help in any way possible." And so, I didn't feel like I was doing anything wrong, but I knew that I wasn't being received in a positive way.

> —Lisa, Cybersecurity Analyst in Cyber Department at a Mid-Size IT Company

My parents packed me up and moved me to Connecticut, and I worked for [a] tech company there. And they had hired seven college hires, so I was one of the seven new college hires that were coming into that team. Of course, I was the only woman, and of course I was the only person of color of the seven.

But what was happening was I was very lonely and alone, and they were very nice people, but they were geeky type dudes, and that was not really who I would normally associate myself with on a personal level. It's fine for work, but now it's like, "Now it's time to go home," and I'm like, "I don't know anybody here. I'm in a place where I have no friends, no family, no church, just nothing." I didn't really want to make an effort to bond but I needed to... but they were playing beer pong and playing video games. It was college-frat-boy type stuff, and I was like, I don't even like beer!

—Kimberly, IT Data Intern

I think I got ignored, one, because I was Black and was the only one in the lab who looked like me. I'm not quite sure that my PI [Principal Investigator] had ever had a Black graduate student before.

—June, Graduate Research Student

Being "the only" doesn't afford us the opportunity to challenge the status quo too often. We are forced to pick and choose our battles as to when to speak up and when to confront bias and inappropriate behavior.

We were diverse in a lot of ways—we had several LGBTQ males and one female. But the only Black female—I was a lot of that in my career. And I would say that early on in my career, I did a lot of giving passes, a lot of giving passes.

—Reagan, Mid-Career IT Consultant at a Small Boutique Firm

Being "the only" doesn't have to be perceived as a bad condition. Part of our coping mechanism is to see the glass half full and instead think, "I am here for a reason," and, "I will show them what a badass boss of an African American female technical professional I am, and I will shine since I'm the only one here doing it."

So, in one of the manager/lead-type roles that I had, I found myself being what I would call "the token." I was the only female. I was the only African American sitting at the table. I looked at it as an opportunity, an awesome opportunity, in management and executive leadership. However, when I looked around the room, I was the token. I was the only one, and at times, it felt very awkward.

> —Kamryn, IT Business Analyst/Project Manager at Multibillion-Dollar Energy Company

The program that I came through was for new grads, and so it was running for about three years, I think, at that time. Maybe, I don't know, 40 to 50 students had gone through the program. And I was the only Black female, was the only person who was not an engineer at the interview. And I actually got the job!

> —Sasha, IT Intern at an International Telecommunications Firm

I became an agile coach. But how I became an agile coach is part of the power of being underestimated. I was one of the most junior people at a part of my practice that would help with sales and business development and that sort of thing, so I learned very early how to do proposals and business development and that sort of thing. And I could process a lot of information quickly and kind of spit it back out as a BA [Business Analyst]. I was up for a promotion at the time as a manager, and I got passed up for promotion the first year.

And I remember I was managing people. I was doing everything that I should be doing, but I didn't get the promotion because they said I wasn't known for anything. So, I was like, "Okay, fine. I'm about to be known for something."

—Vivian, Agile Coach at a Big Five IT Consulting Firm

Many stories of being "the only" can be discouraging, but they just mean that we have work to do. The good news is that it is not all on us. We work collectively to change the spaces that we are in and the people who interact with us daily. Right or wrong, we are often the only ones of our kind as we ascend to higher and higher senior positions in tech organizations. Until that changes, I provide some practical guidance and lessons for you to follow.

Think of it as a journey. Every major journey in all of the best-selling novels and award-winning films start off with a challenge. The challenge is massive, seemingly unattainable, and the quest is not for the faint of heart, although the reward is great. Together we are on a journey, and along the way you will learn more about its challenging aspects. There are also clues to propel you along toward your destination. What's our collective destination, you ask? Why, several more Ursula Burnses as CEOs of Fortune 500 tech companies, of course!

You can choose to leave, not wait, and start your own company, which many of us have done. However, there are plenty of us who have decided to stay, and for us, there is still much work to be done. Think of it as being a Black person in the US in the 1960s. Many were considering leaving the country to reclaim the roots and legacy that was stolen from us by going back to

Africa or starting elsewhere in another country. Some did leave, like James Baldwin, Paul Robeson, and Josephine Baker, but some chose to stay and fight. So, for those who choose to stay and fight, these clues are for you.

Being "The Only" Clues

$1 out of 15 cents

"Making $1 out of 15 cents" is a cultural phrase that is all about hustling. It's about taking nothing and making something. It's akin to the "make lemonade out of lemons" mentality. What it really reminds me of is the Biblical story of the widow who used her last bit of flour and oil to make a small loaf of bread for the prophet Elijah as she and her son were preparing to starve. That little bit of what she had lasted until the famine was over. Being "the only" in corporate America means having to take the little political capital that society has given us, then adding ingenuity, street smarts, high emotional intelligence, creativity, and a strong will to succeed. It's no doubt that we are in drought season when it comes to our representation, but we will not only survive, but we will also thrive! Regardless of our title or degrees of separation from the C-Suite, we are making an impact right where we are, and that impact is helping others.

We are also role-models. I once had a lady work for me, and about a decade separated us, but we instantly connected. I couldn't wrap my head around how this superstar wasn't a Senior Director by this point in her career. After getting to know her story, which consisted of being "the only" at an engineering plant and constantly going from pet to threat like it was a merry-go-round, I understood how her career was not blooming as she deserved, but regardless, wow—what a powerhouse and force to be reckoned with wherever she worked!

Her ability to navigate the "likability zone" with her peers and subordinates was impressive. I was convinced that she could move mountains when I saw her get stuffy, old, set-in-their-ways developers to ask "how high" whenever she told them to jump. Despite her amazing achievements being written off by my senior leadership as anything other than her resourcefulness, she still rose. She didn't work for *them;* she worked for herself every day, determined to make a difference and help whoever needed her.

Managing relationships aside, she was my mentor, coach, and advisor. She pushed me to climb higher because it gave her joy to see another "sister girl" making it. This example of selflessness is a hidden clue. It's not about you; this is a collective effort. Those of you who make it stand on the shoulders of not only your ancestors, but your peers as well. There may be others in other departments outside of product management, engineering, and IT who are looking to us as well. Once we are there, it is our obligation to lift another before we leave; you don't have to wait until you get a C-Suite position, either! I interviewed several ladies who were actively helping to raise up interns and support African American summer students when they were entry-level analysts themselves. The concept of "making something from nothing" will be a lot easier when we all pool our 15 cents together. We will get there faster!

Let Your Light Shine

My father passed away in November 2007. It was a bittersweet time. He had suffered a great deal in his it empowered me to position myself with work enjoyed doing versus work that was given to me. discovered my strengths, and focusing on them was a beautiful awakening.

To let your light shine, you must first find it, then allow it to light your path, allowing you to walk in your greatness and leverage your "only" status to your benefit.

Get Culturally Competent

There is very little probability of success in this journey while being "the only" unless you understand your company's culture and your role in the culture, as well as which rules to play by and which rules to push. Being "the only" doesn't afford any mistakes. Being a student of how your company and department makes hiring, promotion, and stretch-opportunity assignments is critical. Again, as with any quest, there are landmines, pitfalls that will hang you up for long, too long. Ignorance of the landmines, also known as trying to pretend they don't exist, doesn't help. As I've already hinted in the previous section about the need for us to put our 15 cents into a pot to help each other, I will make it plain now. Knowledge of the company politics, the ways decisions are made, and the unwritten rules do not have to be used just for you. You may have made the decision a while ago that moving up is of no interest to you, and that is fine, but help someone else who is trying to get there.

I will never forget a time when I hadn't been with one company long, a few months, maybe. I was 15 years into my career as an IT Manager. At the time, I was preparing a process change that I needed input on from a director in another IT department, so I decided to "drive by" her office to engage in some friendly chat life, battled addiction, suffered a divorce, lost a connection with me, and endured chronic health issues. His life was marked by a lot of sad times and heartbreaking disappointments, but the one thing that he was known for and that we all spoke about during his funeral was his light. He had this amazing, God-given energy about him. When he was at his best, his smile and genuine love for his friends and family extended even beyond his six-foot, six-inch frame, so it was unmistakable to miss the feeling of it with his absence. I pledged that day on November 13, 2007 to a church full of family, friends, and churchgoers that I would let the light that he passed on to me continue to shine.

This clue is all about digging deep and finding your inner light, your passion, and your drive and letting it fuel you. After all, darkness always recedes in the presence of light. Being "the only" will cut deep some days, and many other times, you won't notice it at all. You will have good and bad days; you are human. The myth of the "strong Black woman" will drive you to an early grave if your life is anything like mine—married with children, a full-time job, a passion project, a side-business, and leading a couple of community and professional organizations. Oh, and you are on the company's Black History Month planning committee, too, right? In order for your light to beam, you will need to get crystal clear on your purpose. Regardless of whether this STEM/STEAM career found you or you found it, you are here for a reason, and whatever you ar great at, whatever is easy for you but hard for others that is exactly what you double-down on!

When my sponsor purchased the book *N Discover Your Strengths* for our entire IT departm he had no idea what a gem he was giving me. Fir my strengths enabled me to articulate in job inter and later to my management team what I was gr

Dents in the

before getting down to business. She wasn't there, but what I got was a whole 90 minutes' worth of gold instead! The department's Administrative Assistant (AA) was sitting outside the director's office and popped her head over the cubicle wall like Punxsutawney Phil on Groundhog Day.

We struck up a chat, and she proceeded to download 90 minutes' worth of history of the department into my cerebral cortex. I couldn't process all the names and relationships fast enough, but in the end, she saved me from making a grave mistake with my little process change idea. Instead, I pitched the idea to my supervisor and had my supervisor carry the message. The AA had been with the company for almost a decade and even longer at two other tech companies. She made it clear that her job was to impart everything she knew to "onlys" in hopes of helping them in their rise to the top.

Not everyone has a benevolent, trustworthy soul like that AA who had found me, so oftentimes, doing your own research is required. In Rick Brandon and Marty Seldman's *Survival of the Savvy*, the authors postulate that there is a direct correlation to politically savvy employees and success. They do not encourage unethical behavior such as back-stabbing, lying, or any negative conduct. All it takes is simply lifting your head up from your computer screen or lab table and looking around to start getting curious about your environment. Observation will enable you to perform to the best of your ability and be recognized for the amazing work you do.

We just learned the complexity of how intersecting gender and race can leave us "onlys" invisible, so we must work to be visible in a way that serves us. I love how my fellow ITSMF sister, Monica Brown, repeatedly reminds us in her book *Only One* not to show up playing checkers during a game of chess. Some of us have been out of school a long time, and some of us just graduated and are happy to put down the books and long nights of studying code or writing research papers. I implore you to go back and once again embody the role of student; for this stage of the expedition, you will need a passing grade in cultural competency.

Every person on life's journey needs help. Our travels are long, difficult, and filled with danger—some mountaintop experiences along with some deep valleys. No one in their right mind wants to take a journey like that alone. When the going gets tough, a little encouragement can go a long way, and when you start to get weak and tired, mentors, sponsors, and advisors can step in to offer even more words of wisdom and clues to get you to your next stage. A Barnabas to help with the mission, for my Bible readers.

This is where you come in and we welcome you! Do not shrink back, do not sit in silence, and do not console after a meeting in which your African America colleague was ignored or talked over. Speak up! We want your help and don't mistake help for rescue. We aren't damsels in distress who need saving; we just need someone else who, when they see an injustice taking place or a scale not quite in balance, do something about it. If you have a seat at the proverbial table in an executive or director position, or if you have any political capital to share, there are some guiding practices for you to follow at the end of each chapter as well. As you will learn about the Althea Test, when the organization is doing right by African American women among their ranks, they are 99% likely to be helping everyone, including you!

Active Allyship

The term "ally" has been thrown around quite a bit lately. Many who were silent before are now claiming to be allies to an underrepresented population in this country, naming and proclaiming their allyship to the Black community. Just about every tech company in America threw out "we stand with Black folk" messages and banners on their websites and social media in May 2020.

So, what is an ally, anyway? Historically, becoming an ally has meant forming an alliance with another. Countries can create alliances with each other, like during wartime. Business partners can join forces and become allies to one another during an important bid to win a large customer. Merriam-Webster's Dictionary has added an expanded definition of the word declaring that the term ally is "often now used specifically of a person who is not a member of a marginalized or mistreated group but who expresses or gives support to that group."[6]

I now introduce the term "active ally" from this point on to emphasize the critical nature and distinction. You only get to fly your ally banner when you are actively participating in the elimination of the oppression and unequal treatment of a marginalized and often poorly treated group of individuals. I strongly recommend that you focus on African American women first.

Why?

As you keep reading, you will learn about the Althea Test, used by an expert diversity, equity, and inclusion consultant to assess the diversity maturation

---

[6] "ally," https://www.merriam-webster.com/dictionary/ally.

of any company in just three questions. All three questions center on the treatment of African American women. In essence, start with the group on the lowest rung of the totem pole. The care and feeding of our White female counterparts has been going on since the 1970s in this country. "Time's Up" for us not only means ending sexual assaults and harassment of women, but also specifically including African American women as a focus of diversity efforts. Yes, we must be specific! "Women" as a catch-all does not address the intersectionality of race and class. As author Isabel Wilkerson teaches us in her 2020 book *Caste: The Origins of our Discontents*, race and class win out in this society and are used to define a person's fate much more than gender.

Active Allyship Clues for "The Onlys"

Diversify Your Network

Now that you understand the critical need, where do you start? What is the best way to reach out to our counterparts that seem stuck in the Valley of the Lonely Only?

Step one is to diversify your network. Look for others with different ethnic backgrounds than yours to join for a coffee or virtual meeting just to get to know them. Be intentional about reaching out to colleagues who are outside of your generation, race, and cultural background. If your IT firm is small, then vendors, contractors, and customers are all fair game to connect with to help expand your perspective.

Start simple by introducing yourself and saying something such as, "I've been working here a little over a year and realize I don't know much about engineering, marketing, sales," and then ask them for their time to explain their job. These sessions almost always lead to them discussing their family, children, parents, and a real connection can be made in just 30 minutes! It really is that simple. Imagine how special you would feel if someone from another part of your company didn't want anything from you other than to take the time out of their busy day to learn more about you?

Diversifying your network has tangible benefits for you as an ally. Your reputation as an empathetic and caring leader increases your social capital. Continuing to nurture that relationship will provide informal channels of communication about happenings in other parts of your organization. Lastly, when needed and appropriately reciprocated, diversifying your network can have a direct impact on your ability to complete an assignment by calling in a favor. IT is a relationship-based environment. Many servers have been restarted in the middle of the night based on a favor instead of some VP barking orders. You are now positioning yourself to have friends all over the company by creating a real alliance, also known as active allyship.

Get Curious and Then Listen

Why is reaching outside of your comfort zone so important if you are looking to be a true active ally? You need to learn another perspective. Take a beat and be open to the idea that your experience isn't the only experience. It isn't about you being "wrong;" it is more about the fact that others have different information that has led them to develop different thoughts and feelings than you. Remember, seeking first to understand is our golden rule in leadership, and it is certainly one for active allyship!

"Our findings reveal that structural racism, sexism, and race-gender bias were salient in the women's STEM settings. These experiences were sources of strain."[7] By using the term "strain," the authors mean stress. Women of color are stressed out. It is emotionally difficult to feel like the weight of the race falls on your shoulders in every work interaction, team meeting, and presentation. Constantly showing up to offer a different point of view is emotional draining.

So, what can you do? Ready? It's a two-part assignment. First, ask, then shut your mouth and listen. This is much easier said than done! This requires what us trained coaches like to call "level two listening." Level one listening is internal listening. This is where your thoughts are all about you, your judgement, your feelings, and your experience. This is where you ask someone how you think the meeting went, but you really don't care about what they think; you are just using it as a prompt to share your thoughts about the meeting. You half listen to their response, ready to interject your observations and opinions. Level one listening is focused on yourself and not the other person.

Instead, I invite you to actively participate in level two listening. Level two listening is focused on the other person; you zone in on the person and what they are saying, you observe their body-language, their tone, and their facial expressions. You then paraphrase what you interpreted of what they said, ask clarifying questions at the end of a long pause to make sure you don't interrupt. Instead of your self-talk thinking, "How

---

[7] Ebony O. McGee and Lydia Bentley, "The Troubled Success of Black Women in STEM," *Cognition and Instruction*, vol. 35, no. 4, 2017, doi. 10.1080/07370008.2017.1355211.

does this impact me?" your thoughts are replaced with, "How is what they are saying impacting them?"

Now, here comes the hard part... you must practice! I highly recommend practicing on your spouse, child, sibling, or close friend first. It is not easy. The longer the person speaks, the more intentionally focused you must become. I have found that most people are not used to being listened to at a level two and will fidget, avoid eye contact, and become uncomfortable. Work relationships that have more than one degree of separation, such as age, gender, class, race, and ethnicity, tend to be superficial. When many different degrees of separation exist between individuals, it takes time, effort, and a healthy dose of empathy to build trust. Quite honestly, if you are just starting out as an active ally, this is the best place to start—and you may end here. It is up to you to move into the practice of additional active ally behaviors.

For example, we know that being an active ally takes the courage to speak up. If you aren't ready to vocalize your concerns because you don't feel as if you have enough political capital in the company yourself, then I recommend that the next best active ally posture is to purposefully connect with someone from your newly diversified network and practice level two listening. Being "the only" is hard, and having no outlet to vent some pent-up stress makes it even harder. You are stepping into actively listening not to coach or advise but to be a confidential sounding board that can and will make a difference.

Warning: your newfound level two listening skills may be met with trepidation at first. If the person you are listening to doesn't respond, or if they filter their words and put up a wall, it's okay. Do not be discouraged. Remember, building trust takes time. If you are serious, you will try again, and if all attempts fail, it may not necessarily be personal. Move on to another diverse person in your expanding network. Opening up to someone that challenges your Similarity Bias to be your authentic self is rewarding, not only for them, but for you as well.

Expect Ambition and Prepare for It

If you happen to be in a position of hiring and promoting, this active allyship message is directly for you. You need to know that a large majority (not all, but a high percentage) of African-America women in tech are highly ambitious. If you think about it, it makes sense. We are the ones who fight through microaggressions, stereotypes, and being left out of important study groups. Most importantly, we struggled through needing to quickly catch up from failed high schools that didn't teach us coding fundamentals and failed to encourage us to register for higher mathematics and science courses as well as a lack of guidance counselors encouraging us to take that extra SAT and ACT prep classes on the weekends.

These aspects and more were documented in the findings on African American students attending Predominantly White Institutions (PWI) researched by Daniel Solórzano, Miguel Ceja, and Tara Yosso. We have navigated a rough world through our entire school careers to land at a desk as an entry-level analyst, associate researcher, junior engineer, or Level 1 developer; now that we have "arrived," we are ready to learn, grow, and show what we can do.

When I say we are ready, I'm not playing! Sylvia Ann Hewlett and Tai Green, in their 2014 research from the Center for Talent Innovation, found the following about African Americans who are ready to lead: "However, on one critical front Black and White women are extremely different. White women are skittish about wanting the top jobs in their organizations: they are ambivalent about wielding power. Black women on the other hand are shooting for those top jobs. They are much more likely than White women (22% vs. 8%) to aspire to a powerful position with a prestigious title."[8]

In summary, they discovered that we, as Black women, are three times more likely than our White female peers to aspire for a top spot. We are not the same, and we do not necessarily want the same things as other women. This is the criticality of not lumping all women into one category, sticking a feather in your cap and calling it "diversity programming." By the very nature of the word, women who differ racially are, in fact, diverse. So, prepare yourself for the African American women who want and need sponsorship from you to progress their career. We refuse to be stuck in the Valley of the Lonely Only and will look to you, allies, as a resource to move us along our career journeys.

Breakthrough Tools

- Half of Black women in corporate America are the only ones on their teams or in their departments, and with that comes a great of internal and external pressure.

- Being "the only" can lead to isolation and vulnerability due to stereotyping by others as well as their own self-doubt and insecurity.

---

[8] Tai Green and Sylvia Ann Hewlett, *Black Women Ready to Lead*, 2015, talentinnovation.org/publication.cfm?publication=1460.

- A shift in perspective can enable women who are "the onlys" regain their power and some control.

- Allies are welcome to create space and help break barriers for different races, genders, and/or ethnicities; not to rescue, but to support more inclusive work environments.

Starting Points

Women: Call a trusted confidant tomorrow with an agenda that will strategize on ways to improve or advance your work situation.

Allies: Invite someone who doesn't look like you to a virtual get-to-know-you session tomorrow.

Chapter 2

Bullied at Work

"Do not allow your mind to be imprisoned by majority thinking."

—Patricia Bath,
Ophthalmologist/Inventor/Academic

Remember the angst of middle school? Remember trying to balance who you were while struggling to determine where you fit between the cool kids and the nerds? Mean girls, bullies, pimples, oh my! Corporate America is reminiscent of middle school at times, and just like middle school, IT companies have their fair share of bullies.

Bullying is a workplace hazard and should be listed as a formal cause for medical leave. Bullying is on the rise, taking place at every rung of the corporate ladder and happening far too often. Bullying happens to everyone regardless of gender or ethnicity but is compounded when you are an African American female. Leah Hollis found in her research that the more layers of subordinate groups one belongs to, there is an increased likelihood of being bullied at school or at work, rising exponentially.[9]

The fact that degrees, certifications, and experience aren't all that is needed to advance a career was a major paradigm shift for me. I, along with most every other Black youth in America, have been taught to get an education, go to college, and obtain a degree in order to even be considered for a position. We don't have similarity bias to put us in favor; in fact, almost all the biases (i.e., experience bias, similarity bias, distance bias) put us at a disadvantage. However, there is a notion that a degree, or even better, an advanced degree, will level the playing field or maybe even give us an advantage. Unfortunately, the entrenched power structures in American culture, such as access to networking, prohibit this equal stance from occurring and in fact work to "project exclusionary traditions."[10]

For example, I was asked to mentor a young White male technical analyst to become a project manager (PM). My initial assessment was that he didn't have what it took for the job; his project was failing, and he struggled to execute the advice I was providing him. When I inquired about the need to spend so much time making this guy into a PM when clearly he was better

---

[9] Leah Hollis, "Bullied Out of Position: Black Women's Complex Intersectionality, Workplace Bullying, and Resulting Career Disruption," *Journal of Black Sexuality and Relationships*, vol. 4, no. 4, 2018, doi. 10.1353/bsr.2018.0004.

[10] Audrey Thomas McCluskey, "Setting the Standard: Mary Church Terrell's Last Campaign for Social Justice," *The Black Scholar: Journal of Black Studies & Research*, vol. 29, no. 2/3, 1999, pp. 47-53.

suited for other roles, I was informed that he was the CFO's neighbor and it was a favor. I was hit with the famous line by Tim Gunn on reality TV show *Project Runway*: "Make it work." To be clear, there is nothing wrong about a CFO's neighbor's son getting taken under the wing and receiving some extra tender, loving care; the point is that because we don't exhibit the same characteristics that would remind the CFO of himself at a young age, and because we don't live in the same neighborhood, we have a higher barrier to entry.

This explains the phenomena of Black women being the most educated class in America with a steady increase in degrees from 2014 to 2020 but still only making up 3.8% of managers in corporate America. The percentages dwindle to less than 1% at the Fortune 500 CEO level. Former American Express CEO Ken Chenault said, "There are thousands of Black people who are just as qualified or more qualified than I am," when commenting on a report that found that less than 1% of Fortune 500 companies are led by an African American CEO, and that less than 1% is mostly comprised of Black men.

There are many reasons Black women's careers do not maintain an upward trajectory. Perhaps we left a job, hit a plateau, or even moved to a lower position. However, a toxic workplace ranks right at the top of the negative career impact scale. Bullying makes up the majority of the definition of a toxic work environment. The Workplace Bullying Institute defines workplace bullying as "repeated, health-harming mistreatment of one or more persons (the targets) by one or more perpetrators. It is abusive conduct that is: threatening, humiliating, or intimidating, or work-interference, i.e., sabotage, which prevents work from getting done."[11] "An older 2008 poll on workplace bullying found that 75% of employees reported being affected as either a target or witness." A recent 2019 Monster.com survey found that nearly 94% out of 2,081 employees said they had been bullied in the workplace.[12]

It is time to introduce the school *Stop Bullying* campaigns in the workplace. Since we know bullying is about power and control, you can now better draw the conclusion between a White man with power in the organization against an African American woman in the company with the least amount of power. This power differential makes us a walking target. When you are "the only," being bullied leaves us between a rock and hard place.

All too often bullying is ignored since we don't feel supported in speaking up. We truly believe that our speaking out will make it worse, but just as importantly, it is ignored by those who witness the bullying. Power is a strong force in corporate America, and without it, the perpetrator can't bully effectively. However, aligning against a powerful person can be a death sentence to your career. Risking your career's well-being to try to correct an injustice that isn't happening to you directly takes Herculean strength and courage. To better understand the amount of courage it would take is akin to the same amount of bravery of a young, scrawny kid

---

[11] *Workplace Bullying Institute*, workplacebullying.org.

[12] Bryan Robinson, Ph.D., "New Study Says Workplace Bullying On Rise: What You Can Do During National Bullying Prevention Month," *Forbes*, Oct. 11, 2019, forbes.com/sites/bryanrobinson/2019/10/11/new-study-says-workplace-bullying-on-rise-what-can-you-do-during-national-bullying-prevention-month/?sh=9826c9f2a0d4.

standing up to a big bully in middle school. If that bully is your direct supervisor, taking up the crusade for justice is almost unthinkable and therefore renders everyone silent.

I had been a proud, long-standing member of Zeta Phi Beta sorority for over twelve years. Their values have always aligned with mine in being community-conscience and action-oriented. We don't just sit on the sidelines and do nothing; we dive in to help. That is exactly what I had been doing one late Sunday evening at the beginning of fall semester at the campus of Indiana University-Purdue University.

The undergraduate chapter of the sorority needed a graduate advisor to physically attend their campus-mandated events. I volunteered, knowing it would cause me to work extra on the weekend to compile the new estimates for our department's failing integration program. The initiative was bringing in data from three different systems, and that data included Personal Identifiable Information (PII). I won't bore you with the hoops and red tape this includes, but let's just say it isn't an easy or fast process, which meant multiple project delays—severe ones!

The ladies needing me on campus tugged at my heart strings, so I went, which at first was to be a two-hour event but dragged on into the night. Now, I was filled with dread because I had left my husband at home with our toddler to fend for himself and make dinner. She was an active toddler—and now active tween—so it usually took both of us to keep her occupied and out of trouble for an entire day. We usually tagged in and out when our patience or energy was depleting, but that night, he was on his own.

I remember racing home and pulling into the garage, shocked that I hadn't been pulled over for a speeding ticket to arrive to a haggard-looking daddy and a four-year-old going strong in the middle of the living room playing with her blocks. I felt so guilty, but I had no choice but to grab dinner off the stove and rush to the basement to finish the final program updates for the status report due Monday morning. As my husband was reading a bedtime story, I was hitting send on the final report at 8:31 p.m. Peacock proud, I bounced up the stairs, meeting my husband coming downstairs, and we walked to flop on the sofa to recount our day.

At 9:00 p.m. sharp, my phone rang—it was a company number. I was surprised. It was within our department culture to never call on the weekends unless it was an emergency, and an emergency in my department meant a system being rendered inoperable, but we weren't running anything that was mission critical that couldn't wait until Monday to be fixed. They certainly should not have been calling me! I was Program Manager, not a technical specialist.

When I answered, it was my Senior Director on the phone, looking for the program updates—now.

I said, "I just sent the updates and forecast projections to the team for review. You'll have them by lunch the next day, but I can push the team's review sooner if you need me to."

She said, "Unacceptable. The projections are needed now!"

"But they haven't been vetted yet," I said. "But I can put a draft on the report and send that over."

"Again, unacceptable! Now is now!"

The consequence of not sending the new project estimates were made crystal clear, so to my chagrin, I sent them. My husband could see the distress on my face going from upset to sheer panic. The threat of "if you don't send them now, I will get someone who will" resulted in beads of sweat on my brow, my underarms instantly drenched, and I was pacing so much, I made myself dizzy. My husband kept motioning for me to sit down, but I threw him the "not now" look and proceeded to circle the space in front of the couch.

Ending the call with, "I'll send it over right now," I plopped down on the middle seat cushion, shot a look of dread at my mate of seven years, pushed the rising stomach acid back down to where it belonged, booted up the laptop, accessed Outlook, and hovered over the "send" button. For exactly 60 seconds, I compiled the outcome, ran through all the options and people to call, plus the negative and positive consequences. Then, holding my breath, I hit "send." Not able to speak, I just shook my head in response to my husband's wide-eyed, edge-of-his-seat, arms-open stance, pleading for me to tell him what the call had been about. I said, "Just work stuff. It's over now."

It wasn't over. Three days later, I discovered that my project estimates, which were very incorrect and off by two months, two full head counts, and several thousand dollars, made their way into the company's report to the Board of Directors. My new estimated plan was now the new program plan, and we were expected to deliver it. By the third read of the company status report, I saw red—literally—and experienced my third official panic attack. (Attack one was freshman year during finals week, and attack two was when I was seven-and-a-half months pregnant with my daughter, when my ever-expanding uterus was pushing on my diaphragm and I couldn't breathe.)

That night, I called my mentor and peer of my Senior Director, and in a five-minute, uninterrupted gut-spilling session, I recounted the program estimates, the threat of not sending them, how wrong it was, and how the team was going to hate my guts when they found out this proposed schedule had been approved without their input. For the record, he had the kindest response. He talked me off the ledge, told me exactly what to say in a formal message to cover my tail, and gave me the negotiation steps that I was going to have to perform in order to slowly back out of the plan. I followed his advice step by step; it wasn't easy by anyone's imagination, but we finished, finally getting the extra two months' extension we needed. The afternoon of our team celebration, the Senior Director thanked me and said how valuable my contribution was to the team, and yada yada yada. Two hours later, I submitted a request for a formal release from her team and permission to interview and lead projects in another department—any other department where I wasn't going to be bullied!

    This is only one example of workplace bullying. My story is what can happen in the office, but bullying can also occur outside of the organization's four walls in professional societies, on customer sites, and now more than ever, on social media. About three roles and two companies later, I was a team manager and self-proclaimed "process queen." I announced my department's decision to implement a new process methodology on LinkedIn and expressed how excited I was to be a part of the implementation. A former colleague replied that she was knowledgeable of the methodology and would be happy to grab coffee or lunch and share her learned lessons with me. However, a man in our network openly ranted to our peers that I was "a lamb being led to the slaughter." He commented that this "inept person," referring to the lady offering me advice, who was let go from her last position shouldn't be advising me of anything. He explained that he had been posting on LinkedIn seeking guidance to determine if he should intervene and "save" me from her impending flawed instruction.

Now there was a lot going on there! His comments put him in the role of the white knight coming to save me from impending doom (self-righteous much?), and while nearly slandering my friend and former coworker, he put me in the role of a poor, naïve, innocent, incapable leader. Apparently, I was too incompetent to assess deficient counsel when it was provided.

Do you call the bullies out or let karma take over? Well, in my case, destiny won that round when the white knight did reach out to offer his services to me, in which I was able to decline, commenting, "Gee, thanks, but little ol' me just wouldn't know what to do with your help." A polite "go fuck yourself" is always in order in these types of situations.

Frontline Tales

I was an IT manager at a manufacturing company, actually... and then upper managers, which were all male, White male. They would undermine decisions that I made on the floor and even go behind my back and talk about the decisions that I made. And so, there was a lot of the bullying there.

[And the staff were horrible as well.] ...there was just a difference to the rules. I was bullied by the employees themselves, some of them, being called "colored" for the first time ever in my life. White males refused to listen to me as far as advising them during the shift and refusing to go to the departments I had assigned them to go to.

—Dominique, IT Manager in Rural Indiana

I've been bullied. I worked for a medical device —orthopedic device—company. And I was responsible for performing system validation studies for new IT rollouts. The director who I reported to, she would just cuss me out on the daily. And—yes. She used actual profanity. She actually had attacked me, came into my cubicle yelling. I was on the phone, and the person who I was talking to heard her cussing me out. ...She just didn't like the way I worked. Nothing I did was good enough for her.

—Dominique, IT Validation Specialist

I was in charge of change management processes for the department. The team had to request changes to me. If they were outside of what should have been approved, I rejected the requests. My supervisor wasn't happy with my rejections. He said I can go a lot, lot further if I just say "yes" more than no.

The same supervisor berated me on my general performance weekly. And he would go on to ask me what I thought that I'd done well for the week. I'd go through a litany of things that I accomplished, things that I felt good about, things that I had done, and then he said, "That's funny you should say," and this was weekly for about six months. And he would say, "What I want to let you know about is all the things that I feel you've done wrong. So, I think we're on two different pages of understanding what right and wrong is because I feel like all the things that you said that you did well, I feel like you did them wrong. And you're an embarrassment to the team." And that was weekly.

So I did make him aware that I felt bullied. Some of the words that he used were threatening. Some of the tactics of standing up over top of me while I told him the things that I did right—he would say that he talked to other people that were in the meeting who said that you did terrible in the meeting and that you didn't know what you were talking about. He would talk about how I needed to be more professional. And I'd ask him, "What do you mean by 'professional?' Is it the way I dress? Is it the way I —what is it?" And he would say, "I just don't like the way you conduct meetings." So, I never could get anything substantial that I could develop on.

And just before I gave my notice [to leave], he said he wanted to put me on performance [performance improvement plan, or PIP], and HR said, "There's nothing wrong with her performance."

—Laura, Senior Program Manager at Multinational Engineering Company

It was [about] approval. It was approval for certain monies and documentation. And at the time, one of the customers needed to hide money from their particular government. So, they overpaid for a product, and they wanted me after three years to release that money back to the government, back to that customer, because they were basically hiding money from an audit.

> I said, "I will not do that." And I told them they'll need to do it the right way or find somebody else to do it, and they said, "We will absolutely do that, and we'll take those responsibilities away from you." . . .and they did. [I was forced out shortly thereafter.]
>
> —Alicia, Validation Lead

Like these ladies, sometimes we are forced out of a role for not succumbing to other's expectations of how we should do our jobs or break the law, as in the last example, but oddly enough, other times, they want to force us to stay.

> I gave my notice and was immediately met with, "Well, where are you going?"
>
> I said, "Excuse me. I'm giving you my notice."
>
> They said, "Well, if you don't have another job, you need to stay until you finish this project."
>
> And I said, "It was my professional courtesy to give you a two-week notice, and that is what I'm doing."
>
> And so, they literally tried to bully me into staying.
>
> —Paulette, Agile Coach,
> Minnesota Tech Agency

> We were talking, and I think [he was] just getting really annoyed with my dominance, to be honest. And I think before I joined the team, he probably was the biggest voice of the team. And I come in with my big personality and all my ideas and all my

energy, and I just think that was off-putting. And it wasn't like I was doing anything negative or wasn't trying to be inclusive. I think some people just—they internalize it differently. And I think that was happening there.

Later, my supervisor called me in to review the feedback that I had provided for someone on my team. We were in a meeting, and he just started yelling at me. I was told that I was too emotional when I gave [my employee's] feedback, and I had to change it. So, even though I recommended him for the highest raise you could get, I had to change the [written] feedback. There was no option. It was, "You have to change it!" It was the most demeaning experience of all—I mean, I've had lots of experiences, but that one was—[look], I'm a leader. I gave feedback for an employee, and I was told that I'm too emotional for it. And after that, I was stripped of a lot of leadership opportunities.

—Stephanie, IT Manager for a Banking Institution

So, I'm talking, and he starts interrupting me. . . going over me. So, I turned to him, and this is the exact thing I said. I said, "I'm not finished." And I turned back, and he said, "How dare you speak to me that way?" . . .[Frank] is the kind of person who has a lot of bark, but if you bark back, he withers."

—Paulette, Agile Coach

He would make threats, like, "Well, if you don't do X, Y, Z, then I'm not going to support you in this manner." This particular person actually ended up taking over our lab and being the PI [Principle Investigator]. And then, during that time, I had gotten this major award to have funding to get a PhD. And he tried to manipulate me and tell me, "Oh, well, you can't leave this lab. You wrote the grant based on research in this lab, so you have to stay here." And I didn't want to stay. I wanted to try to publish the work I did in that lab before I moved on because that's the proper thing to do, and he wanted no part of it. So, I left anyway. Then, once he became on the tenure track, he reached back out to me. He wanted to [publish my work] several years later, even up until 2016. [I said], "Well, you know, I've moved on."

—June, Principal Researcher

He wasn't my supervisor. He was the head of another department. We were all standing in his office answering questions about an issue. He says, "You should have sent me this information," and [I] was like, "I did send it to you," and he's like, "Do you want me to put your head to the computer and show you I didn't get it?"

—Monica, Head of Engineering at Multinational Industrial Corporation

Bullied at Work Clues

Write. It. Down.

Document. Document. Document. Being bullied in the workplace puts you in a defensive position. This position is uncomfortable at best and physically anxiety-producing at worst. Physical symptoms of stress can take a toll. Unlike schoolyard bullies who threaten to beat you up after the three o'clock bell on the playground, work bullies can pounce at any time of the day via a nasty email or an unannounced swing by your desk. The pop-in bully is worse since you don't know when or where to expect them to strike. They can take your blood pressure from 120 to 210 in eight seconds.

To get some control back, I recommend first taking a very deep breath and writing the incidents down. In one of your brightly-colored personal notebooks, simply write down every time this person triggers you. The date, time, and circumstance. After they leave, write it down quickly to get the negative energy out of you. Revisit later when you are ready to think about it or talk to a trusted counselor or mentor about it. The trick is that you do eventually have to come back to process the occurrences—don't leave them unexamined in the back of your notebook for too long.

> There are choices. I worked for one company for numerous years, and that's why most people don't do that. So, most people are used to potentially finding a next opportunity. If they can't do that, then you need to document, document, document, document until you're ready to talk to someone formally about a complaint, because otherwise, you're going to get interrogated anyway, so you better have the dates, what was said, who said it. I mean, that's the best advice I can give. But otherwise, it's like, leave.
>
> —Tina, Value Management and Six Sigma Manager of IT Services

Another option would be to keep an email folder with a nickname for the bully that only you would know. Every nastygram, threatening message, and inappropriate remark can go right into the folder. Thanks to email rules, you can set a rule in Outlook or Gmail that will automatically send notes directly from them into "their" folder you created to be read when you are ready. These are small steps to regaining some control.

Call 'Em Out!

What do we know about bullies? They are insecure! They are scared little boys and girls on the inside who appear to have very rough exteriors. Although I've met a fair share of bullies who had a nasty bite and were extremely vindictive, you don't want to get caught in the wrong bullies' crosshairs, so you must observe the bully in multiple interactions before you move forward with this clue.

Calling a bully "a bully" might just be enough to stop them in their tracks. Bullies come at various levels of intensity. Level One bullies tend to have a very low level of emotional intelligence. Daniel Goleman coined the term emotional intelligence which is essentially one's ability to be aware of their own emotions and how to handle others' emotions empathetically. Level One bullies are anything but self-aware. They may genuinely not know how brass and harsh they are being. In saying, "Frank, stop bullying me into making the decision to fast track the project," replacing the word "bullying" with "pressuring" may be enough to shock them back to reality. I would expect them to be defensive, snarky, or dismissive, but if they are Level One bullies, they will stop with the direct high-pressure tactics enough to at least give you a breath.

Level Two bullies will crank up the fire. They will move to passive-aggressive maneuvers after being called out. Level Two bullies know that their intimidation tactics work. They successfully achieve their goals by yelling, demanding, controlling, and scaring people. Level Two bullies require a deeper level of strategic countermands, so start by observing who they bully and when. Observe their behavior while they are around their own supervisors—is it Dr. Jekyll/Mr. Hyde, or do they bully in front of their manager as well? Do they only bully subordinates or peers also? Do they bully mostly within their own department and people outside of their department love him/her, or is it the opposite? Is the bully targeting just a certain group of people, like women or African Americans, or worse yet, are you the sole target of the bully's offensive behavior? The answers to these questions will help you develop your plan of protection.

I call it a "plan of protection" instead of a "plan of attack." Attacking will make Level Two bullies worse; therefore, your primary priority is to protect yourself from further stress, plus mental and emotional abuse. Once you have observed the bully in multiple settings and in relation to others, the next step is to take this information to a trusted advisor or mentor—someone more senior than you who can remain objective, level-headed, and keep your information confidential. You will create your plan of protection and discuss how to execute it. Exploring what-if scenarios, role-play, and breathing exercises are all techniques you can utilize to, once again, gain control over your situation.

Now, what to do with Level Three bullies? This is no longer kids' play once you encounter a Level Three bully. They demand and command total control and tend to be the worst micromanagers you can imagine. I've only seen (from afar) one true Level Three bully. He was a VP-equivalent in the upper echelons of IT senior management. He struck fear in the hearts and minds of all around him, while Level One and Level Two bullies cowered as he approached.

If you crossed him or did something stupid, like let a vendor accidentally push untested code to production, causing a mission-critical application to go down in the middle of the night, it was off with your head! Well, they didn't let him behead anyone, but you were demoted or moved to a different department far, far away, never to return. His name became a verb, so when he said, "Jump," everyone asked, "How high?"

Battling or just working for a Level Three bully is not a laughing matter. They hold your hard-fought career in their hands, so doing exactly what they say, when they say it, means living on pins and needles day and night. Our fierce African ancestry would tell us to stay and fight and press on until the victory is won. However, the cost to our hair falling out, constant upset stomach, and/or hair-trigger angry outbursts leave our friends and family feeling the pain, even if we are thinking we are "handling it." So, once you realize that you are dealing with a Level Three bully and the toll it is taking is too much to bear, it's time for your final clue.

Call 911

"Calling 911" means that it is time to call for real support. You need assistance—professional assistance! This could take the form of an attorney, professional counseling, or a practicing HR professional (outside your company) to craft your best possible outcome.

To graduate from being forced to work with a Level Three bully, you'll need to revisit all that documentation we just spoke about. Before you go busting down the door of a HR representative, I strongly encourage you to share your encounters via documentation with an attorney or HR rep that you trust that doesn't work inside your company. You will need legal advice about how best to achieve your goal, which at this point is probably a request to move teams or departments if you wish to stay at the company. Most of the cases that I have personally witnessed are those in which the wrath of the bully is too severe, and the target resigns.

Should you choose to stay with the company but want to put some distance between you and the bully, you will need to finesse your way into crafting a solid argument. For instance, moving is not only in your best interest but in the organization's best interest as well. You will need to solicit help from the bully's peers when the bully in question is your direct supervisor. Anticipate that a Level Three bully may not like you or want you around any more than you want to be around them. However, if you request a transfer, unless it was their idea, it will be met with resistance. Therefore, the sage advice of a legal professional, HR rep, or someone versed in the laws and regulations of your state is crucial. Knowing your options will allow you to gain control, mount a solid plan of protection for yourself, and move forward with confidence, regardless of the outcome. If you choose to fight, fight. If you choose to leave, leave. Whatever you do, it will be your choice!

Active Ally Clues for Coworker Bullying

Codependent No More

My mom was a counselor during the time when the term "codependent" hit mainstream. Although not quite clinically accurate, I always associate codependency to mean "enabler." Whenever I see a person or group of people allowing bad behavior to continue, I think to myself that they are codependent with the other person. Codependency is not just linked to drug or alcohol addiction. Enabling can happen at work as well.

If you are seeking to be an active ally and witness the behavior of a Level One or Level Two bully, it is time to speak up. Someone other than the target calling the bully out on their behavior is much better than only the recipient of their torment doing so. In her article *How to Stop Workplace Bullies in Their Tracks*, Christine Comaford encourages allies to not be silent and to create a culture in which bullying is not tolerated. With as high as 75% of survey respondents in Comaford's article reporting being bullied or seeing someone else bullied, these results indicate that bullying is quite prevalent, and bullies know that their treatment of others can only go on if allowed.[13]

Bullying is very bad for business. All those recruiting and hiring efforts to locate and on-board diverse talent go right out the window if an employee, especially an African American woman, leaves because they are being bullied. Belonging to a majority population in the company, you as an ally have more political capital than a Black woman, especially if she is "the only" in the department or on the team. You are in a better position to call attention to the bully's behavior as unacceptable.

---

[13] Christine Comaford, "How to Stop Workplace Bullies In Their Tracks," *Forbes*, Mar. 12, 2014, forbes.com/sites/christinecomaford/2014/03/12/bust-workplace-bullies-and-clear-conflict-in-3-essential-steps/?sh=74247fb57912.

For example, when I challenged my coworker, who was just as tired of our department bully, to speak up for me, he was shocked. He was stunned at the idea that he had some power, certainly more than me, to improve our situation. He eventually nodded in agreement as the idea had settled that he had some influence to help.

If the thought of standing up to a bully gives you a physical reaction, imagine how the person who is targeted feels. This is the point in which you get to earn some active ally stripes. The option of you not saying anything to improve the situation doesn't exist. Tactics that you learned back in the Valley of the Lonely Only, such as empathetic listening, are a start. Telling the target of the bullying that you see her pain, don't agree with how she is being treated, and would like to discuss how you can work together to minimize or eliminate the bullying is a good first step. If your morality beeper starts going off, then, depending on which level bully you are working with, it may be a good idea for you to consult with a trusted advisor or mentor before speaking up to ensure that your advocacy will be effective.

I will not minimize the risk. It is precarious to your career to give up some political capital in order to stand up for a teammate or to a manager bullying their staff. Prepare yourself for the best way to be a supporter, which may include garnering help from others on the team so you aren't the lone person speaking out.

Accountability Rules

A culture of bullying is difficult to change. Individual contributors and people in the minority typically do not have the political capital to make effective or lasting change in this area. If you are in the majority and/or in a senior-level position, you have the power to step up and enact change. If you see bullying by a peer, another Director, or VP, for example, you can change the culture or enable it. Don't think too hard about this choice. Please cease your enabling behavior.

Permitting a culture of bullying means that you are turning a blind eye to the behavior. Your silence speaks louder than the bully's yelling or berating if they are in one of your meetings and you do nothing to stop them. The tips below can be easily carried out in the setting of a physical meeting space but can also be translated to virtual meetings as well.

Shifting to virtual meetings presents a wonderful opportunity to establish or re-establish ground rules. Introduce team agreements that include respecting and killing the discouraging behavior of certain individuals talking over others. Comaford outlines specific steps that can be taken to combat bullying from persisting under your watch:

- Clarify what appropriate meeting etiquette is, and ensure it is honored. [In other words,] enable team accountability.

- Instead of holding your anger in and feeling as frustrated as others, [stop avoiding it and] deal with this issue directly, modeling leadership for the team and showing them a safe, respectful, collaborative work environment is required at the company.

- Instead of creating an opportunity to let disgruntled parties communicate their grievances directly and interface with HR, stop allowing them to come to you to moan and complain, [only] for you to not do anything.[14]

Remember, the name of the game here is to become an active ally. Your assignment in this situation is to lend your power to someone who doesn't have much. Confrontation, action, and outcomes-based solutions are what is needed to help someone overcome a harrowing experience such as being bullied at work.

Breakthrough Tools

- Bullying at work is more pervasive than we think, and it can happen to anyone. But when you are "the only," it accentuates the problem while simultaneously allowing it to progress.

- Bullying is about someone leveraging control over someone using the power that they have and nothing else.

- Assessing which type of bully is targeting you will help determine which strategy should be implemented to provide the most help.

- Staying in a toxic work environment is not worth your mental health. Devise an exit strategy by utilizing your professional network.

---

[14] Comaford, 2014.

Starting Points

Women: Observe the bully in other settings and around other people at various hierarchies. Assess what type of bully you are dealing with in order to devise the most appropriate protection strategy. Document each and every bulling incident.

Allies: Observe the bully in question in other work settings. If you have enough political capital, speak up; if not, acknowledging to the target that you see their experiences as well as tracking the bully's behavior will absolutely help.

Chapter 3

Deep Cuts

"My mother raised us to think that if we worked hard, and if we put our end of the bargain in, it would work out okay for us."

—Ursula Burns

Constant and consistent neglect takes us into an area that results in some deep wounds. Deep cuts occur when you are ignored, your idea is taken, or you are consistently passed over. Would you rather have to deal with your coworker openly berating you with questions in every meeting, constantly sending nastygrams, and telling you to your face that they don't think you are good enough for your position, or the passive-aggressive comments, backbiting, and saying they will have one of their team members contribute to your project, only then to give that same staff member a high-priority assignment, rendering them totally unavailable to you? Neither choice would be great, I know, but you only get to choose one or the other. This game-playing should remind you of the middle school culture we just left, but the subtle, mean-spirited conduct cuts deep, warranting its own platform.

Over the years, I have decided that I prefer the in-your-face folks. I personally have a bit more respect for those who are bold enough to tell me how they feel about me to my face. Every now and then, the bias is not always unconscious. The overtly racist and sexist comments remove the extra time for mental hoop-jumping. I now have less time to spend on wondering if a team member, supervisor, or staff member is giving me the cold shoulder or being rude and dismissive of me because I may have personally offended them in some way.

Overt forms of racism and/or sexism enter the atmosphere usually in the form of a joke. In 2012, Hall, Everett, and Hamilton-Mason outlined in their research the type of stress Black women endure at work and how they cope. They recorded one subject of their study recount the following exchange with her coworker:

"One day I was asking him (White coworker) something and he was talking, and he was like, "Would you hold on, it ain't February." I said, "Oh, hell no, he didn't say that," and so I was like, "Don't you ever say anything like that to me ever again," and then he really got like, "I don't understand why she got mad."[15]

These types of racist remarks are usually followed by the phrase "I was just joking." The joking defense by the perpetrator now puts the responsibility back on the offended party to let it go, not make a big deal about it, and if it is escalated to others' attention, she is seen as "not able to take a joke." This is the

---

[15] J. Camille Hall, et. al., "Black Women Talk About Workplace Stress and How They Cope," *Journal of Black Studies*, vol. 43, no. 2, 2011, pp. 207-226.

number one reason why I and many other African American women are so serious at work! Letting our guard down, getting too comfortable, and moving into the comfort zone with our peers too often results in a slip, opening the door to a racist or sexist joke which reveals the not-so-hidden stereotypes.

We must keep our guard up to balance the thin line of moving into the "likability zone" but not into the "comfort zone." The risk of allowing White people to get too comfortable and start making racist or sexist jokes is simply too high. It puts us in the position of either a) ignoring it, which is a HIGH price to pay because that means more jokes will continue and exponentially increase the psychological trauma we put on ourselves for feeling like a "sell-out" and not standing up for ourselves and our people, or b) being forced to confront it like the respondent above. Her response was direct and unquestionable in the fact that her coworker's comment was offensive and racist, but now she's labelled as the dreaded "angry Black woman."

Whenever we stand up and confront racism and sexism head-on, the "angry Black woman" stereotype will be triggered. So, to avoid the trap, we turn into contortionists to maneuver around the comfort zone while trying to stay in the "likability zone," if we have been fortunate enough to even make it there. Hall, Everett, and Hamilton-Mason concluded in their study that additional research is needed to understand the complex intersectionality of race, gender, and class for African Americans to understand the full picture of what we experience in the workplace.[16]

The lack of intersectional analysis can result in an incomplete picture that excludes crucial perspectives

---

[16] J. Camille Hall, et. al., 2011.

and gives little attention to why disparities, such as those between White and Black women, in earnings, advancement opportunities, unemployment rates, and other areas continue to persist. For example, conflicting dichotomies of stereotypes for African Americans exist versus those facing African American women specifically. African Americans, especially men, are characterized as being lazy, lacking work ethic, and needing to be watched carefully to ensure they are not slacking or producing poor quality work. African American women can be stereotyped as either the mammy, the Jezebel, or "angry Black woman."

The mammy archetype is happy at work, enjoys working, and will work for free because it is just what she does and it's what she's good at, versus the Jezebel, who is promiscuous, lazy, and will do anything to satisfy her carnal urges. While the "angry Black woman" or sapphire archetype takes herself too seriously, is difficult to work with, and will always openly confront prejudice, so you better be careful what you say around her to avoid being labelled or "told on" when your jokes get too off-center. The initial assumption is that I am lazy and need to be watched or else I won't do my work. Or that I will work nights and weekends with no expectation of increased compensation, promotion, or reward. Or you don't want to work with me because I'm too serious and not fun. You can quickly see as an African American woman entering a new work environment, I have no idea which perception and stereotype I'll be battling, and the battle begins even before the on-boarding and orientation.

To combat these long-held American caricatures of African American women, we employ a coping mechanism called identity-shifting. I like the term introduced by Charisse Jones and Kumea Shorter-Gooden in their 2004 book, *Shifting*, much more than just saying that we are "code-switching." Code-switching is the practice of moving back and forth between two languages or between two dialects or registers of the same language at one time. Nordquist states that "[it] is common to find references to Black speakers who code-switch between AAVE [African-American Vernacular English] and SAE [Standard American English] in the presence of Whites or others speaking SAE. . . .For a Black person who can switch from AAVE to SAE in the presence of others who are speaking SAE, code-switching is a skill that holds benefits in relation to the way success is often measured in institutional and professional settings."[17]

For example, if a White female coworker responds that she is tired because she was up all night with a sick child, a typical response would be, "I completely understand" versus if a Black female peer relayed the same information, a typical response would be, "Feel ya," with a possible "girl" or "honey" tacked on the end, depending on how close the relationship is outside of work. As an African American woman, my natural inclination is to speak in a relaxed, causal, informal manner to build close bonds, relations, and demonstrate equal status. However, I observed that my White peers, no matter how casual or close, would still say "I understand" to each other. Any other response could trip a stereotype. As the pressure mounts to be credible, educated, and competent, I and other Black women in these corporate work environments absolutely cannot risk being too casual—not even once—so our mental filter goes up and stays up all day to avoid any slips, trips, or falls into AAVE terminology.

---

[17] Richard Nordquist, "Learn the Function of Code Switching as a Linguistic Term," *ThoughtCo.*, July 25, 2019, thoughtco.com/code-switching-language-1689858.

However, Dickens and Chavez report that the language filter is not the only change African American women choose to make while working. We change our hair, dress, name pronunciation, and other characteristics to assimilate to the point where our entire identity has been altered. "As a result, Black women may use different coping strategies, such as identity-shifting, in the workplace to protect themselves against experiences of discrimination, invisibility, and marginalization. Because of the daily engagement in identity negotiation, work-life can become psychologically exhausting and stressful."[18] To clarify, we must be White enough to our White community but Black enough to not appear to be a "sell-out" or "Uncle Tom" to our Black community.

To be a "sell-out" in the Black community is essentially going too far in your identify shift to where you now believe the changed identity is who you are all the time. Think Hillary Banks in *The Fresh Prince of Bel-Air*, part of the show's comedy centered around their young "fresh" and "hip" inner-city cousin's dress, language, and "down" interactions at their Beverly Hills prep school. In the Black community, it would have been acceptable for Hillary to dress, act, and speak like a "Valley Girl" at her predominately White school, but the expectation would have been to drop the act at home or around other Black people. If you don't switch back to Black, a red flag is immediately sent up, like a red flare on a distressed ship, signaling to other Black people that you may have forgotten who you are by playing this corporate America role for too long.

---

[18] Danielle D. Dickens and Ernest L. Chavez, "Navigating the Workplace: The Costs and Benefits of Shifting Identities at Work among Early Career U.S. Black Women," *Sex Roles: A Journal of Research*, vol. 78, no. 11-12, 2017, pp. 760–774, doi.org/10.1007/s11199-017-0844-x.

Additionally, the concept of intersectional invisibility suggests that individuals with multiple subordinate identities (e.g., Black women) do not usually fit the prototype of their respective subordinate groups and thus will experience subtle or invisible forms of discrimination.[19] This means that choosing which identity you portray in your work life is critical. The wrong selection can have dire consequences and leave you walking an even thinner tightrope. The difficulty of navigating between a gender minority group in America's tech industry and the other ethnic subordinate group of being African American leaves the average African American female tech employee completely spun out on a daily basis. Despite the subtle consequences of being visibly invisible, there are tangible penalties as well. According to Frye, left unaddressed, these different narratives and workplace dynamics will continue to have real consequences for Black women's earnings. They also reveal how discrimination and stereotypes become entrenched in workplace structures and practices.[20] Bottom line, ignoring the more visible minority in the room doesn't render the corporate silent; it screams volumes to the person being overlooked.

---

[19] Valerie Purdie-Vaughns and Richard P. Eibach, "Intersectional Invisibility: The distinctive advantages and disadvantages of multiple subordinate-group identities," *Sex Roles: A Journal of Research*, vol. 59, no. 5–6, 2008, pp. 377–391, doi.org/10.1007/s11199-008-9424-4.

[20] Jocelyn Frye, "Racism and Sexism Combine to Shortchange Working Black Women," American Progress, Aug. 22, 2019, americanprogress.org/issues/women/news/2019/08/22/473775/racism-sexism-combine-shortchange-working-black-women/.

I'm not sure when the term "STEM," later revised to "STEAM" to include the Arts in Science, Technology, Engineering, and Mathematics, became a household (and now politicized) word. When I was growing up, it was science—just science. Everything from biology to computers was lumped under that term. So, I thought the need for more Americans to focus on STEM and the need to support, encourage, and nurture more women and minorities in technical fields was new. Well, it is not new at all! In fact, a landmark study published in 1978 detailed the experiences and expertly laid the execution plan to support, encourage, and nurture not just women, but minority women, in what we now call STEM/STEAM:

> Malcolm, Hall, and Brown were commissioned by the American Association for the Advancement of Science (AAAS) to understand the issues of minority women in science for the purposes of advancement. The AAAS organization created an Office of Opportunities in Science:
>
> - to increase the number of minorities, women and the handicapped in the natural, social, and applied sciences;
>
> - to increase the kinds of opportunities available to these groups;
>
> - to increase the participation of minorities, women, and handicapped scientists and

engineers in policymaking advisory and managerial positions.[21]

You can deduce that the AAAS was a real-deal ally from the start. They invested time and finances to create a conference dedicated to learning about and understanding us as Black women! A group of 30 ladies, just like the ones I interviewed and whose stories you hear throughout this book, shared their experiences in science with the organization. From their stories, the intention is set to create new policies, enact supportive changes, and create a more inclusive environment for African American women. The preface of the report contains this excerpt:

> "There was little information available on the status of minority women in science and virtually no literature that would advise institutions on the nature of the problems or the remedies. The AAAS therefore decided to arrange a small conference of the women themselves to find out exactly what the problems are, and in what respects they are similar to or different from those of majority women scientists, minority male scientists, and all other scientists. It was to define and illuminate those questions and to receive the advice of minority women scientists that the conference was held in December 1975 with support from the National Science Foundation. The immediate and enthusiastic response of minority women to the

---

[21] Shirley Mahaley Malcom, et. al., *The Double Bind: The Price of Being a Minority Woman in Science*, report no. 76-R-3, American Association for the Advancement of Science, 1976.

announcement of the conference is evidence that the meeting was long overdue.

This volume is the report on that conference. Its policy and program suggestions can serve as a guide to the public agencies, educational institutions, professional associations, and funding organizations. Perhaps its larger contribution, however, will be in helping the rest of us in science to understand more clearly and fully the situation of those who are excluded from the mainstream. As we become more aware of their perceptions we can move more readily toward equitable solutions."[22]

I can't say exactly what has transpired or hasn't in the 45 years since this conference took place. I'm sure history will tell us that some gains were made, some areas were improved upon, and some women benefited from the work of the AAAS. However, it is abundantly clear that the vision Dr. Carey had has not yet been realized. If it were, *Dents in the Ceiling* would not exist, and the thirty ladies I interviewed would have had less content to share about microaggressions, bullying, and not being visible at work.

The intersectionality of race and gender can hit three times as hard for us. Sometimes it is a racist comment, other times a gender-based comment, and every now and then, you get the double-whammy of a two-for-one. I love the way the authors of *The Double Bind* summarize this paradox from their conference attendees: "It must be noted that several instances were

---

[22] Ibid.

recalled which could have been brought about by race or gender bias or both. When this situation arises, it becomes difficult if not impossible to determine which "ism" is in force. In such a case, it does not matter whether one is being hit with the club of sexism or racism—they both hurt. And this is the nature and the essence of the double bind. I, along with female, minority, science forebears, am left wondering which one (racism or sexism) is more offensive, but I'll let you decide."[23]

The summer of 2005 was a very busy time for me! I was planning a wedding! I'd met a fraternity brother of Phi Beta Sigma Fraternity in graduate school and pursued a long-distance relationship. To put it mildly, a long-distance relationship between Connecticut and Indiana was not working for me. Don't get me wrong—I enjoyed the growing frequent flier miles and weekend trips to fancy restaurants and museums. We were both in IT, so the paychecks supported our young, single, and fun lifestyles, but the emotional roller coaster was increasingly harder to ride.

After a bit of a false start (another book, another day), we began serious wedding planning that left just under six months to pull it all together in time for Labor Day weekend. It was a big wedding with 150 guests, most of which were out of town since neither of us were from Indiana. It allowed me to practice my budding project management skills on the biggest project I'd ever managed at that point in my career, because at my job, I was the youngest in our entire department. I was the youngest person on a floor of 80 statisticians, data scientists, quality and validation specialists, coders, managers, and directors—the youngest! My wedding

---

[23] Ibid.

became my "likability factor." This was the safe topic on which I could connect with everyone in order to achieve the right level of being likable.

As I popped up from my cubicle and headed to the ladies' room, I saw my supervisor walking down the hall from my left while our chief Stats Product Owner, a Senior Director equivalent, was walking down the hall to my right. The pace of their walk resulted in their meeting right in front of my cube. While I smiled and nodded to them both, hoping to slide by my supervisor as I wiggled toward the ladies' room, he embarked on a bit of small talk, so I paused to engage. It was clear that our Product Owner (and I, for that matter) had other places to be, but my supervisor kept talking. Now, this was about a week prior to my wedding, so my coming nuptials were mentioned, and my supervisor remarked, "Well, guess we better start looking for another project manager for your project. Angel will probably come back pregnant from the honeymoon and leave us."

In another situation, I was screaming inside but didn't know who to turn to for help. I was a brand new employee and project manager. I had just reviewed the stats of my team members. When I inquired about the $50k pay difference between my male and female colleagues doing the exact same work, I was informed that he had more years' experience than her. I wondered, *A few more years equates to a $50k pay differential?* I continued to push.

"Then why isn't he in a higher job position, a more senior level role that would align to his salary range?" I asked. "After all, he is making more than me, and I am his supervisor!"

The response: "He really likes this job, and he is really good at it, so we keep paying him to keep him."

In summary, I was told that the man in question, who reported to me, gets to receive pay increases for doing the same job quarter after quarter with no increase in responsibility or expectations. While I didn't agree with the reason I was given, I shifted my focus to my female colleague who was clearly making less than the man in question. I started this phase of the conversation by exploring what she had to do to reach his level. I was told, "Oh, it will just take time." There was no special development plan required; the direction provided was "just time." At the rate her annual merit increases where projected, it would have taken her 20 years to catch up! That's a hell of a lot of time!

There were other examples of women being treated differently at this company. Lisa, for example, was hired to advise the department senior executives and team leaders on a new management style. Instead, she was given a room with no windows, sat by herself, and was explicitly told to "stay here while we send people to you." During a lunch encounter, I witnessed a tense and uncomfortable exchange between Lisa and Monica, who was acting as the "representative" of the department's senior executive team. Lisa was inquiring about attending the next staff meeting in which she was told to "just stay here." Well, she stayed in that room for about two months longer and then left. A few months later, a male consultant was brought to perform the same role and was seated in the middle of the team, given full access to all systems and files plus authority to make changes to team structure and processes.

At a previous company, I remember it being my first week on the job with my coworker Richard, who was about six-foot-three, a husky, middle-aged White man who, by all accounts, was extremely good at his job as a senior IT business analyst. Richard had a deep, rich, booming voice that was hard to miss, especially in an open-work area where there were no walled offices or even cubicle desk barriers. So, when Richard told a joke about the Black IT developer he used to work with and how funny this "Black fella" was and the "dumb mistakes he made," I heard it loud and clear. I not only saw him look at me and another African developer in the room and say, "Oh, no offense." I also saw red!

Three months into the assignment as a Release Manager on a newly formed team, I was having an intense conversation with the head Validation Coordinator, Oliver, who was in our U.K. office, and our program's chief Product Owner, who was based in the in the U.S. with our team but that day happened to be on the phone. We were arguing over the content of the Release Plan and what additional information I needed to include before he would approve it. My stance was that this was version 1.0 of the document, and version 2.0 would come soon, but to meet the mandatory quality deadline, what I had was enough. He disagreed and wanted more information on our processes. Processes we hadn't even created yet, since it was a new team. The Product Owner, Leslie, could have easily stayed in Switzerland, playing her stance neutral and mitigating the disagreement. Instead, she came down on his side, which left me in a two-against-one battle.

By this time in my career, I had more than a decade of experience under my belt. I also had enough coaching and mentoring to be more confident and push back with data-driven, articulate arguments. However, emotion was starting to creep in because our supervisor and the program's head sponsor, Ronald, had obtained agreement with the Validation Coordinator's boss that the current content was acceptable, and we should move forward. This meeting, I thought, was going to be a brief chat about the formatting of the document and approval sign offs, but it instead turned into a battle royale.

Tensions were high and frustration abounded, and I decided to cut the conversation short for fear that explicit name-calling would commence. Not reaching a resolution, I ended the dispute with a "I'll touch base with Ronald after he lands and get in writing that we can move on." (Ronald was known for sending out directives and hopping on a plane before the details were clarified.) However, after I made that final statement, they didn't realize that I was still on the conference bridge. Leslie and Oliver thought I had hung up. The line, however, didn't disconnect, and I overheard Oliver telling Leslie how funny it was to let me twist. He said, "They are lazy anyway," and that it "won't hurt her to work a little harder." Could the "lazy they" he was referring to mean Americans, or women, perhaps? No, he was talking to Leslie, very much an American woman. Oliver meant Black people. I not only left the team and that department but resigned from the company about two months later.

Being ignored occurs blatantly and subtly. For example, the State of Black Women in Corporate America 2020 report from Lean In found that when a Black woman succeeds, people often attribute her accomplishments to factors outside of her control, such as affirmative action, help from others, or random chance.[24] For example, colleagues might say things like, "She only got the promotion because she's Black," or, "She was lucky to close

---

[24] *The State of Black Women in Corporate America*, Lean In, ,/2020, Lean In.org/research/state-of-black-women-in-corporate-america/introduction.

that sale." This reinforces a damaging stereotype that portrays Black women as less talented and competent than her peers. When these comments go unchallenged, they can prevent Black women from receiving the credit we deserve for our hard work and achievements. When our contributions, such as to the project, to the new technology service, or to the sale are explained away, we are being ignored.

Frontline Tales

> I remember someone trying to call me by my official name. And I go by a nickname. So, the person [tried] to call me by my full formal name and I said, "Yeah, my name, right? It's okay. You can call me..." and I offered my nickname again. I know it sometimes is very hard to pronounce my full name. And the person completely ignored me and still called me by my formal name [which they mispronounced]. "Well, I appreciate being referred to as," [and I again I offered my nickname]. "I guess that should be better." And they didn't apologize at all. They just carried on. I think my name has been pulled apart, shredded, murdered, and whatnot. And I think the one that used to get me is when I pronounce it and that person just dismisses it.
>
> —Alicia, Senior Agile Coach

> I'm in meetings where we would be having open discussions with the leadership team, and I may have an idea or suggestion, and literally, the next sentence would be another peer of mine saying

the same exact thing and being heard and that reception being there. And I would be like, "Well, that's what I just said, right? Was I the only one listening?" I'm like, "Where is the disconnect?" Whether it's true or not, they put you in boxes, and they just don't listen.

—Tracey, Director of IT at Electric Manufacturing Co.

I'm trying to think. It just happened last week, or I think it was earlier this week. I mean, stuff like that happens all the time. You get so used to it, it's kind of like it rolls off you. But I was in a meeting. I made a statement about something—I can't even remember exactly what it was—saying, "I think it should be. . ." And then someone on my team said, "I agree with [her]," [and the person said my name]. And then someone else says, "Well, we're going to go with what he said." So, it's like, wait a minute. Originally, I said it!

—MaryAnn, IT Manager in Data Sciences

All of the things that you're supposed to do as far as getting funding and all of that, I did all of those things. And then people ignore you. I think I got ignored . . . because I was Black and was the only one in the lab that looked like me. And that needs to change because there are Blacks that are getting PhDs. There are Black women that are getting PhDs. So, somebody needs to figure out, well, what's happening to them after they're getting their PhDs? [They are being ignored.]

Dents in the Ceiling 103

—June, Senior Researcher

You all are just flushing money down the drain. It's not my money personally. You all are giving me a check. I say thank you. You want me to sit here and wait, I say, "Okay." But it still was bugging me. And even though I'm asking these questions and I'm making these suggestions, no one's listening. Like, no one cares. Finally, 18 months later—I probably might have been singing this tune 12 months ago—we finally got to the point where the project has been placed on a permanent hold.

—Kamryn, Senior Systems Analyst of Energy Company

[So, basically, I was] given the assignment to go to an international meeting. Before leaving, I asked my supervisor, who had asked me to go to this meeting to represent [the] team—when [I] asked them for an agenda and [I] asked them for logistical specifications... I never got a response.

[I ended up having to ask a coworker for the agenda so I could reserve my flight and hotel room, but once I got there, I didn't even know where the meeting was being held. I had to ask around. I was never provided a travel itinerary or agenda. I just had to figure it out for myself.]

You have to know me. I'm not one—first of all, I'm not intimidated, and I'm not one to be ignored. So, the only way that they could actually block me was lack of information. I've had the occurrence where I was going to several international meetings over several periods of time with no information, no agenda, no map... no nothing. I had to [go] from one person to the other, and [finally] somebody goes, "Do you have the [meeting] agenda?" "No?" "Here it is." And that was the pattern [to get what I needed].

—Charmaine, Senior Government Aeronautical Staff Member

As a young construction engineer at my first job. . . one day, I went in and the manufacturer's rep was in there talking to the contractor and his team. And when I walked up, the contractor was trying to make eye contact with the manufacturer to basically tell them who I was. But you have to realize I'm a young Black female with a hardhat on. I had a hardhat. I had a hardhat and a clipboard with a pen.

So the manufacturer dismisses me, just like that, and he's telling the contractor how to falsify the test because the equipment was faulty. As the contractor is trying to nudge him to let him know who I am again, he's dismissing him. When he finished, I politely let him know who I was and what the ramifications of what he had just said were going to be.

—Maxine, Construction Engineer

I was in a meeting in Houston with an international partner. The international partner spoke to the interpreter, not to me. Now, I'm sitting right there. He insisted on speaking to the interpreter. You know the difference? Instead of looking at me, talking to me, he's talking to the interpreter?

So the guy says, "I will never forget what you told him." He said, "You told him, 'First of all, I am the person who is responsible for this. That is the interpreter. Second of all, what you have said is a waste of my time. I have not been given directions from my management to do what you're asking me to do. And this is what you need to understand.'" And I remember this part of it. I told him "no" in about six languages. Now, listen. His reaction was he responded to me in English! And as he's speaking to me, I am rising from my chair. I tore up the agenda and told him he was wasting my time and that I had to go out and do bigger and better things.

—Charmaine, Senior Government Aeronautical Staff Member

[There was a lot of work, and I couldn't do it all myself, but my senior leader said that it was just going to be me and my supervisor, no one else, and he expected us to work together.] And it's to the point we [my supervisor and I] would go on trips; he wouldn't tell me he was going. This is my boss. I would get on the plane, and he would be on the plane. He wouldn't talk to me... it was just so bizarre.

—Paulette, Agile Coach

I've been with this company here over five years now. So, since then, I've been telling them over and over, "You're behind. You're behind like ten years. You're behind. We were doing this back then. You're behind. You're behind." So, I just found out that they hired a consultant company and paid this consultant company I don't know how many thousands of dollars to tell them that they were behind 10 to 12 years. When the information came to me, actually, we were meeting about this probably right before the holidays. And there's this, of course, White man. He comes in, and he's like, "Yes, we hired blank-blank consultant company and reviewed our procedures and our operations. And they have determined that we're behind 10 to 12 years." I went, "You don't say? Yeah. How long have I been barking this? I've been barking this over five years, and okay. You went and hired somebody and paid them thousands of dollars to tell you?"

—Jalesaa, IT Business Consultant at Financial Services Company

I feel like myself and another Black woman, we've been labeled the "angry Black women" because we're the ones in projects or in different meetings saying, "Iceberg ahead. Get into the raft. The rats are coming up," you know what I mean? Like from Titanic. And they're just looking at us like we're going, "Wum wum wum wum," that we don't mean anything.

But let a White person say the same thing, "Oh my God, I think there's a big iceberg, Becky," now it's, "Oh my gosh, she's so brilliant. We need to get in the lifeboats."

—Char, Senior IT Associate

Deep Cut Clues

The Mute Button

You have a voice. You didn't misplace your voice in the middle of the night, nor did it disappear the moment you walked into your first team meeting. Stop listening to people telling you to go "find your voice." Your voice is not lost. Your voice, however, was muted. If you are speaking in a meeting and ask a question or suggest a new idea only to be met with silence, then your voice has been muted. The person(s) with more power in the room can physically hear you but are not listening to you.

This clue focuses on gaining the power to take back the remote control and unmute yourself instead of waiting for someone else to do it for you. The first step in commanding attention is increasing your confidence. Lack of confidence stems from being unsure and not being the expert, so you feel like what you're saying or asking will reveal your ignorance on the subject. This is completely a mental game that we play with ourselves. No one is an expert on every subject. Even the leading winner on *Jeopardy!* (rest in peace, Alex Trebek) gets a question wrong now and then. Stop holding yourself to a higher standard than everyone else. I get it, trust me! You already feel as if all eyes are on you and that you are representing your entire race with every question or comment, so you have a strong desire to be right. Release that pressure today, right here and now. You don't always have to be correct, but you can frame your inquiries to allow room for additions instead of deletions. There is nothing wrong with being known for asking good questions. Entire careers have been made from people who know how to ask just the right question and at the right time.

For example, "help me understand," or "have we considered," or "can we pull on that thread a little more, please," and my favorite that I learned in a coaching class, "I'm telling myself a story. . . is that accurate? What might I be missing?"

When looking to pose a recommendation, there's no need to announce or negate it before you've said it. I cringe when I hear myself say, "this might not be right, but. . ." or, "call me crazy, but. . ." Just make the recommendation and include two to three points as to why your recommendation is valid. Silence? No one agreeing or disagreeing? Then it's time to call people out. Not in a malicious way, but by inviting others to add to your comments, such as, "Jake, what are your thoughts about my recommendation?"

Does that cat have Jake's tongue? Was Jake feeding his cat off camera on the Zoom call and missed what you said? No worries. Just respond with a little humor and move on to someone else. Confidence in your question or suggestion will give you the extra edge you need to solicit feedback when you were first ignored. A great tip that was shared with me when someone else takes your idea is to restate your idea again in the form of a thank you. "Thank you, Jim. That helped to clarify my comment of. . ." and restate your original idea here. You must pick and choose your battles, of course, but standing up for yourself will greatly increase your chances of snatching back the remote and getting off mute.

Practice Makes Perfect

Voice, tone, inflection, and non-verbal body language, such as posture and hand gestures, are variables that make up how well you are heard and can help you avoid being ignored. You will need feedback from someone that you trust, like a mentor, to provide you feedback on how well your message is being articulated. Do not feel bad unless they start adding drama class as an elective to Computer Science programs. These are skills that you may not have ever learned or had an opportunity to practice.

Have an idea in mind that you are preparing to share at the next team meeting? Need to present a project update or share data from your research? Practice prior to the session. I am a professional speaker. I get in front of meet-up groups, have been a part of panels on company main-stages, and speak at conferences in person and virtually, and I still couldn't quite get heard at my team meetings. I even tried props to get my idea across, to no avail. I leveraged my executive coach to listen to my presentation from beginning to end. My content was solid, but my delivery was rushed and didn't focus on the points that the audience cared about. I was sharing content that I cared about. Thank goodness for my coach calling out my blind spots. I practiced at least two more times, even recording myself, which is always very "cringey," as my daughter would say. Once you have researched what to say and have practiced how you will say it, your confidence will increase.

Caroline Goyder, in her TEDx Brixton talk on speaking with confidence, teaches us that while practicing, you should focus on controlling your breath. One of the ways to regulate your breathing is learning to activate your diaphragm properly. She posits that diaphragm control will help center you and shift your physical breath from your chest or throat downward, providing a grounding or centering effect, which forces you to slow down and speak confidently and clearly.

One of Goyder's most impactful statements is, "We breathe our thoughts."[25]

Making your presence distinct, impactful, and effective for your goal. Practicing how you show up in the room, getting feedback from a trusted source, and watching yourself talk are all great ways to make your presence known. Once you have done it successfully and see a different outcome, you will be encouraged to try again. As they say, "practice makes perfect," so keep developing on those wins even if you face a setback. So, let's move on to your next clue to help you increase your chance of being listened to at work.

On the Campaign Trail

As I sit here writing this book, it is election time here in the United States. Candidates from Senators to police captains are on the campaign trail sharing their ideas for improving the community, reminding people of their track record, and, of course, tooting their own horns about how they are the best person for the job. Oh, did you think that interviewing for the job stopped once you were hired? Oh no—that's when the campaigning begins.

There are a boatload of studies about women who feel uncomfortable sharing their accomplishments. Well, no duh! We are socialized from birth to downplay our accomplishments, not be a braggart, and, certainly for minority women, we are acutely aware of what bringing too much attention to ourselves could mean. In school, I wasn't a hot shot at mathematics, especially Algebra, so my mom got me a tutor (several, in fact)

---

[25] Caroline Goyder, The surprising secret to speaking with confidence, Nov. 25, 2014, youtube.com/watch?v=a2MR5XbJtXU.

and placed me in Sylvan Learning Center. My F slowly rose to a C+, and on a big eighth grade math exam, I pulled a B. My reward? Being accused of cheating by the teacher. What brought my mom to the school in the middle of day was that it was not only myself accused of cheating, but the only other Black student in class was as well. Needless to say, that parent-teacher conference did not end well for the teacher.

The message to me that day was that buckling down to work hard in order to show how smart I am will not be believed. I told myself that I just wasn't good in math and never would be. Fast forward to my freshman year in college, and thanks to a Teachers' Assistant named Noel, who apparently was a math whisperer, I was rocking A's all semester in Calculus!

The moral of the story is that our environment can give us negative feedback, or we interpret it negatively, which reinforces us to shrink. Diminishing our greatness is not an option. We are made in God's image, and He has given every one of us unique gifts to share. Your purpose, your calling, your passion—whatever you choose to call it it's yours, and no one gets to dictate how you use it. The moment you realize what you are meant to do, you get to tell, share, reveal, and market your talents. There is finesse in how you launch and execute your self-promotion campaign, so a professional coach is strongly recommended, your career progress can easily be stalled by not having one.

The biggest benefit of getting clear on what makes you great and telling others about it is that you can now search for an organization that is a match for the work you wish to do. Saying that I was nervous to leave a company I had been with for over a decade is an understatement. However, once I was clear on the work I wanted to do, and when the company was clear on the limited opportunities it had at the time for me to operate in my gift, the decision to leave got easier. I left with a smile on my face and in search of a round hole instead of trying to fit my round peg in the square hole that I found myself in far too often. It was not rainbows and unicorns searching for the right organization, but in hindsight, it was well worth the effort.

Finally, now that you have your passion and know how to evangelize yourself, you must learn the environment where you are working. Your company culture, your direct senior leadership team, and, most importantly, your immediate supervisor's political style directly impact your marketing plan. Authors of *Survival of the Savvy*, Brandon and Seldman began their explanation of two main political styles in corporate America by quoting Lord Philip Chesterfield: "Learning is acquired by reading books, but much more necessary, learning. . . is only acquired by reading men. . ."[26]

The two political styles the authors describe are the power-of-ideas style and the power-of-person style. The styles are on a continuum, with power of ideas on the left and power of person on the right (not related to conservative versus liberal U.S. politics). The power of ideas is all about a focus on getting things done right, meritocracy-based decisions, results, and ideas speaking for themselves, and there is welcoming of feedback and learning. The power of person is more focused on positional (think hierarchical) power, image and the importance of managing perception, doing what works versus what is right. The power-of-person style believers have more private agendas and make more relationship-based decisions.

---

[26] Rick Brandon and Marty Seldman, *Survival of the Savvy: High-Integrity Political Tactics for Career and Company Success*, Free Press, 2004, pp. 24.

Imagine the frustration of being a power-of-ideas person working for a power-of-person style. Neither is right or wrong, just very different styles that can lead to miscommunication, distrust, and a tense working relationship. For example, after a contentious performance review cycle, we had a young manager who did not receive good reviews from her coworkers or other managers in the department. Despite her poor performance, she was still given a promotion and put into a high-achievers program. Her supervisor shared the news that she would be given the promotion. She was also given the constructive feedback. As a result, she launched into an adult temper tantrum.

After the display of bad behavior, promotion and entrance into the high-performers' program should have been an instant reversal. But the VP had already submitted her name, and rather than admit his folly, being a high power-of-person political style, his image and self-management took precedence over what I considered to be doing what was right. The incident of the VP not wanting to look bad among his peers and supervisors made it clear that he and I were on opposite ends of the political spectrum. I personally asked him to reconsider his decision, and he dropped his head and said, "I can't." At the time, I thought he really meant that he wouldn't, but after learning and understanding the different corporate political styles, I realized he really couldn't.

It is important to note that not all power-of-person styles are brown-nosers or yellow-bellies who do whatever it takes to climb the ladder; those people are great at building relationships, reading a room, and getting work done. Wielding your political savvy for good and not evil can be quite beneficial and having someone that subscribes to a different political ideology than your own is a great find! Use them to see a different perspective, craft a plan that appeals to both styles, and best yet, remember that these people excel at self-promotion and impression management. For example, one of my sponsors has a power of person style. He constantly reminds me to put myself in the position of those being impacted by my process changes and craft my message to address the impact on them. We don't always see eye-to-eye, but him reminding me of how I impact others helps me to fill a blind spot. Remember, they may not be your "cup of tea," but you can learn a lot from them to help with your personal marketing campaign.

Self-Advocacy and #IamRemarkable

A healthy sense of confidence is the most critical building block toward self-advocacy. The term self-advocacy came to rise in the 1950s and 60s around those who were disabled and the need for them to be seen and heard regarding laws and policies to improve their work and life. Now, you will see the term show up often in the educational space, especially when we as parents must encourage our children to advocate for their needs. Self-advocacy absolutely applies to work as well. Your supervisor (and it may feel like it at times) is not 100% of the equation when it comes to your career progression. They may be at that company, but not in your career. You must make a mental shift to decide who is accountable and responsible for your career.

Once you absorb the fact that it is you, then self-advocacy becomes a concept that you can embrace and learn to practice. Start small by meeting with the person who owns a process or system that is posing a barrier to your progress. Build your stance from a position of why they should care to help you, and add a couple of points of how it will be beneficial to meeting their needs if they spend time or a portion of their budget helping you. For example, a lady I work with needs a different report from our project management system. After submitting her enhancement request via the normal request routes and not getting the priority, she advocated for the report in terms of business impact and benefits. Two weeks later, she had a draft report sent to her, which was one small step to improve her proficiency and productivity.

A wonderful method I learned for practicing self-promotion in a non-threatening while aligning with my collaborative spirit comes from the #IamRemarkable campaign. #IamRemarkable is a global Google initiative founded by Anna Vainer that strives to empower everyone, particularly women and underrepresented groups, to celebrate their achievements in the workplace and beyond. To date, #IamRemarkable has reached 130,000 participants and 800 companies across 130+ countries with the help of 6000+ facilitators. I attended an #IamRemarkable workshop, and I am inspired to keep it going! One Friday a month, I kick off an email and text message thread that starts with #IamRemarkable with my coworkers, friends, and family. I state all the things I did that make me remarkable.

Examples from the #IamRemarkable campaign include:

- I am remarkable because I won my company a $2M customer.

- I am remarkable because I led a remote project planning session with almost 50 team members across multiple countries.

- I am remarkable because I completed a new Agile Certification.

- I am remarkable because didn't curse my mother out today.

- I am remarkable because I asked my supervisor to lead a training session, and he said yes.

- I am remarkable because I resisted the bait from a guy who normally gets on my nerve.

- I am remarkable because I got to interview a famous author.

- I am remarkable because I completed a Request for Proposal (RFP) on time, which was the largest RFP my company has ever submitted.

The #IamRemarkable campaign serves two purposes. First, you realize how wonderful you really are! You do so much great work every week and are a corporate America warrior, so every now and then, you must take a step back to realize all that you've done. A quick five-minute exercise to list your biggest hits will certainly boost your confidence. Secondly, the #IamRemarkable exercise allows you to share your achievements and invite others to do so as well. This communicates to your colleagues the work that you've done (feel free to include work friends outside of your immediate group—the more exposure, the better) and creates a culture of celebration. You don't have to wait to improve your culture; start today by sharing what makes you remarkable, then ask others what makes them remarkable.

There is a style of IT delivery known as Agile, which incorporates pillars that rest on high collaboration, feedback loops, incremental and iterative delivery, and high innovation that I study and teach,. Therefore, the political style of power of ideas resonates greatly with me. I suspect that if surveyed, a great number of African American women subscribe to this style of politics as well. I recommend that everyone take the political savvy assessment, determine where you are on the continuum, find a friend on the opposite end of the continuum, and use the data to launch your personal marketing operation.

You already have the clue of cultural competency in your toolkit, and now, you'll pull it out to use it once again. In your observation of your culture, which political style is used most often and also valued and rewarded more—the power-of-ideas style or the power-of-person style? Once you have the answer to that secret, you will be on the path that will get you the limelight you so rightfully deserve.

Active Ally Clues for Neglected Team Members

Chivalry was Not Bad

During medieval times, chivalry was a strict religious, moral, and social code for knights. Chivalry was all about courage, honor, courtesy, justice, and a readiness to help the less-advantaged. These are all qualities that are good and clearly necessary when seeking to be an active ally. But has the art of being polite been lost? What is wrong with simply acknowledging that someone made a comment and thank them for their contribution? No one ever said that you had to agree with the comment or suggestion.

Practicing being polite is an amazing start at being an active ally. The very next time you are in a meeting and someone makes a comment or suggestion, do these two specific actions in this order:

Action 1: Thank them for their comment. Including their name is a bonus.

Action 2: Use "and" in your response to their comment if you have a rebuttal or something to add.

Additive phases such as "and," "in addition to," "expanding on," and "that makes me think" are phases that validate the person's comments and demonstrate that you not only heard what they said but listened to it and appreciate their contribution. These acts seem small but have a profound effect. Saying someone's name during a casual interaction makes the person feel recognized. If someone says your name with a smile on their face and in agreement, this will start to create a feeling of belonging. You have the power to create this same feeling for someone else.

The key to this active ally clue is that you are modeling behavior for others to follow. Remember, if you are in the majority population on your team, you unconsciously set a standard of behavior. You are a role model for your team and peers. Use this hidden power to affect a positive outcome. The payoff has the potential to be huge. If you are successful and get other leaders to join in your style of thanking others for their ideas and questions and not ignoring them, you could slowly create a culture of belonging. You can help create a place where ideas are valued, everyone's voice is acknowledged equally, and no one is negated, where each question or thought is expounded upon, not pushed aside. In "Radical Change, the Quiet Way," Debra Meyerson provides an excellent example of how one person can be a catalyst for positive change.

> Brad Williams was a sales manager at a high-technology company. During a meeting one day, Brad noticed that Sue, the new marketing director, had tried to interject a few comments, but everything she said was routinely ignored. Brad waited for the right moment to correct the situation. Later on in the meeting, Sue's colleague George raised similar concerns about distributing the new business's products outside the country. The intelligent remark stopped all conversation. During the pause, Brad jumped in: "That's an important idea," he said. "I'm glad George picked up on Sue's concerns. Sue, did George correctly capture what you were thinking?"
>
> With this simple move, Brad accomplished several things. First, by indirectly showing how Sue had been silenced and her idea co-opted, he voiced an unspoken fact. Second, by raising Sue's visibility, he changed the power dynamic in the room. Third, his action taught his colleagues a lesson about the way they listened—and didn't. Sue said that after that incident she was no longer passed over in staff meetings.[27]

---

[27] Debra Meyerson, "Radical Change, the Quiet Way," *Harvard Business Review*, October 2001, hbr.org/2001/10/radical-change-the-quiet-way.

20 Questions

In the Valley of Lonely Only, I requested that you sit on your hands, shut your mouth, and listen. I asked you to practice active listening at Level Two, force those feelings of judgement to the back of your mind, and curb the urge to ask too many questions. Well, now that you have advanced your listening skills, assisted an "only" out of her valley, and helped her navigate being harassed and/or being ignored, you've earned some trust to ask questions. We encourage your inquisitiveness now. It is important to establish trust first before launching into a game of "20 questions" with an African American woman. Most of us are accustomed to answering uninformed inquiries about our culture, music, and hair, but you will get a better, more informed, and more honest response after a foundation of trust has been laid.

I was a summer intern for a company in Hamilton, OH. No, not the city of Cincinnati, but the very rural, White community of Hamilton in the late 90s. I was the only African American woman on my team. There was an African American man on my immediate team and other minority interns dispersed throughout the company, so at least I always had someone to sit with at lunch. Think of Beverly Tatum's book *Why Are All the Black Kids Sitting Together in the Cafeteria?: And Other Conversations About Race*. We were exactly that group of Black students all sitting together in the middle of an almost all-White corporate cafeteria, laughing and talking about music, what crazy inappropriate questions we were getting from our teammates, what cool new technology we were learning, and, of course, our weekend plans.

The company sponsored a summer golf tournament in which departments competed to raise money for charity. It occurs during the summer, so interns were invited. Each company I interned with—and there were three—all did this exact same golfing event. Well, during this particular summer internship, I landed on the team with Mike and John. John was a slender, six-foot tall manager who didn't say much but was very fair and objective in his responses. He kept to himself but was always professional, polite, and responsive when I needed anything. Mike was younger, 5' 8" with a stocky build. (I think he said he played football in college.) He was an extroverted and unpredictable manager. Although a hoot at company parties, during work he was prone to fits of outbursts if something wasn't going his way.

Well, Mike was certainly volatile on the golf course as well. He threw a fit every time he missed a swing. It was really John and Mike playing while I drove the golf cart. When John made it to the green, he would encourage me to put it in the hole. I would miss, and he would pop it in the hole on his next turn, and we'd move on. Even with my misses, John won. I survived the heat of the hot July sun and missed a driver to the head that Mike had thrown as well as a family of ducks waddling down the lane as I tried my best to steer the cart and not have Mike's intoxicated butt roll off the back. I hardly survived the car ride back to the office, however. John was driving with Mike in the passenger seat, and I was in the back, Lean Ing over to catch the wind from John's window since his air-conditioning wasn't functioning properly.

Mike turned up the radio to listen to his favorite early 90's tunes, and this is when the song O.P.P. by the hip-hop trio Naughty By Nature was popular. Mike sat up, cranked up the dial, and started to blast O.P.P. like a

Black teenager rolling down Peachtree Street during Freaknik in Atlanta. My eyes widened, my heart began to beat fast, and my palms were sweating as Mike sang and bumped along to the music. John had his hands on the steering wheel at ten and two and was looking straight ahead. I tried to catch John's eye in the rearview mirror, but his eyes were glued to the road. I cringed waiting for him to get to the chorus, and he sailed through, but by the second refrain, he abruptly turned the music down, swung over his left shoulder, put his pudgy, sweaty arm on the front seat (there was no middle console like they have in all the cars now), and inquired, "So what does O.P.P. stand for, anyway? I've always wondered."

A thousand thoughts ran through my mind in that 20-second period as I stammered to think of a response. *Is he testing me? He knows but is wondering if I am crass enough to say it? Is he genuinely inquisitive and wants an honest response? Is he drunk and doesn't know what he's asking me? Does he really not realize how inappropriate it is to be asking an intern, a Black female intern, this question? Why is he asking me? Why not John?*

I asked and answered my own questions in seconds and responded with the best, most appropriate, and professional response I could muster, and, to this day, I am still extremely proud of my reply: "Oh, Mike, just like the song says. . . it stands for 'Other People's Property.'"

He said, "Oh, okay. I always wondered if that was the real meaning. Thanks!"

All the air escaped from lungs from fear that he would press and not accept the politically correct response. I rested my back and head on the seat and glued my eyes to the road ahead, just as John did until we made the journey back to the parking lot.

I was so grateful to see another intern returning at the same time, so I thanked them for a great afternoon and ran to her car. Ashley asked how my afternoon was, and still in shell shock, I just replied, "Long and hot." She scoffed and said, "Yeah. Mine, too." I didn't tell Ashley about my car trip with Mike or the other interns the next Monday at the café. I kept that incident to myself.

I share my car ride with John and Mike in hopes that I can provide insight for my new active allies. Understand that questions are fine, but context, tone, and history of your behavior all build trust. The amount of trust established can either put you in the active ally category or ignorant White man/woman who is just "being funny acting." Funny acting in African American culture is what my grandmother would call some "folk have to be fed from a long-handled spoon." This means you can't be trusted, and you might bite while someone is trying to "feed" you, or help you. Know that when you ask questions, the filter is going up, and the guard wall is stacked high until we can ascertain your true motives.

Tone Deaf No More

Willful ignorance is no longer being tolerated. You can no longer claim, "I didn't know." The murder of George Floyd propelled the plight of the African American population to the forefront of your cerebral cortex. The old way of doing things is no longer acceptable. The old way of getting things done at work, and the overall work culture is shifting. We are shifting to be more inclusive of different people, diverse thought, and dissimilar styles.

All this change commands attention. If you continue to operate as you always have, you run the risk of being perceived as tone deaf. This means that you are viewed as not having a pulse on the needs of your staff, being an exclusive leader, and, over time, falling behind and being labeled as anything but an innovative, forward-thinking leader.

Embracing diversity, equity, and inclusion is good for business. I personally was able to secure a meeting for my company, who is on the path to having a more gender- and ethnic-diverse senior leadership due to my membership in an African American network of highly influential Black business leaders. My mentor introduced me to the CEO of a technology company looking for the software my company produces. Therefore, my company now has access to a population of business owners they wouldn't have if not for my networks. These business owners would not have had my company on their radar to purchase from if I hadn't had a seat at the leadership table. If diversity and inclusion (D&I) efforts seem too soft for you as a hardline businessperson, look at it as another channel for business development, lead generation, and tapping into new markets. In other words, more customers and more money.

I encourage you to consider adding D&I to your portfolio of skills as another currency to your political capital bucket. Altruistic motives are great and can translate into tangible benefits for your company. In early 2020, I wrote a LinkedIn article titled "Work Allies: Black and White Make Green," with the White male executive in mind. It said:

> Anyone today in mid-to-executive management knows the seven principles of the Leader Bible, and our prophet Steven Covey tells us that one of those key leadership principles is to seek win-win outcomes. Your willingness to intentionally help high-potential minorities is the trifecta! It

helps the company save recruiting and retraining dollars. The Society for Human Resource Management (SHRM) reports that, on average, it costs a company six to nine months of an employee's salary to replace him or her. That means $40,000-$56,000 just walked out the door (not to mention their knowledge). Your direct support provides a more inclusive environment; we all know that if a child has at least one friend at school, their odds of staying in school and having an improved well-being increase. The same could be said for minorities in corporate America's workplaces. In short, you can help someone! Last but not least, demonstrating actions that are out of your job description to go above and beyond while saving the company money go directly onto your performance review.[28]

If you are following my logic, your direct facilitation of protecting company assets—in this case, human capital, or knowledge—and stewarding company benefits can have a secondary benefit for your personal career as well. Learning, being in touch with your fellow man (or in this case, woman), and helping someone else allows you to leave a legacy. After all, you hope to retire at some point, passing your baton to someone else, so you can sleep in, vacation, and improve your handicap, but those are not everyone's aspirations. You may subscribe to a more individualistic ideology that includes "every man or woman for themselves" and "pull yourself up by your bootstraps," but that is unlikely if you are reading

---

[28] L. Angel Henry, "Work Allies: Black and White Make Green," Jan. 9. 2020, linkedin.com/pulse/work-allies-black-white-make-green-henry-mlis-mba-pmp-csm-popm/.

this book and learning how to be a better ally. But if you are still skeptical, managing your personal brand to not be seen as "the old guard" is needed as more and more companies make D&I competency a requirement for the C-Suite. Organizations like to hire entry-level associates who are well-rounded and didn't just graduate with a 4.0 GPA, but who also performed community service or achieved something unique during their time at their university. Similarly, companies will start to look for senior executives who cannot only run a Profit-and-Loss center or generate revenue or bring in new customers; they will also expect you to foster a culture of belonging and inclusion for everyone in your care.

Breakthrough Tools

- The intersectionality of race and gender offers a complexity that creates a gap, allowing diversity programs to miss Black women.

- Stop saying that you've "lost your voice." It is not lost! You still have your voice; look to increase your confidence instead.

- Learn to market your own accomplishments and become your own cheerleader.

- Practice makes perfect. It takes practice to learn when to speak up, what to say that will add value to the conversation, and how to package your ideas, so keep trying. In time, you may need a new team, department, or even a new company, but know that your contributions are valuable.

Starting Points

Women: Start sharing your ideas. If you've stopped, reengage. If ignored, bring the conversation back to your idea by thanking the person who expanded on it, then elaborate a bit more on your thought with an example.

Allies: Look and listen for idea appropriation in your meetings and give credit to the originator when it happens. Oftentimes, these situations can be corrected with mindfulness of others. People "riff" off one another's ideas all the time; simply saying "thank you" to the person who started the brainstorm is a great start.

Chapter 4

Maid Service

"When people come to you with problems or challenges, don't automatically solve them. The first thing you're doing is actually putting it back to them and saying: 'What do you think we should do about it? How do you think we should approach this?'"

—Shelley Archambeau,
American Businesswoman and Former CEO

I was chatting with a fellow IT colleague at a conference, and we jokingly made the analogy of how working in corporate America is like working at the Big House or mansion on a plantation. When you start to explore this analogy closer, the laughing stops. Oftentimes, African American women can be made to feel like the wait staff regardless of title. If I can just make it to Program Manager, I will gain respect. If I can make it to Manager, then I will have the authority to make changes. Now, I need a Director position to have true authority in my company. Well, once I have VP behind my name and a whole staff working for me, the constant fight and proving that "I deserve to be here" will be over, right? Wrong. No matter how high the climb, the number of team members under your authority, your title, or your certifications, the fight to not be treated as the help persists.

There is nothing wrong with taking meeting minutes, fetching coffee, and setting up the meetings. The issue comes when those tasks are not fairly and equitably distributed. Being mistaken for the intern or junior analyst is easily a consequence of African Americans looking younger than their chronological age, a blessing socially while a curse in the boardroom.

I was getting restless after about eight years at my first company. I was looking for more responsibility and to have an opportunity to manage others formally, not just their work. After two phone interviews, an all-day round of in-person interviews, paying for my travel, hotel, and per diem for me and my husband, I was turned down for a Director role because I "just didn't look mature enough," but out of fear that the senior project managers wouldn't respect me if they had to report to me, so I didn't get the job. Another recruiter suggested that I pull my hair back, wear smaller earrings, wear glasses, and be sure to wear an "older" suit with pumps to my next interview. Believe it or not, I did—I tried my best to look like my mother just to be taken seriously. Nevertheless, despite my list of degrees and certifications, title, years of experience, and grandma-looking attire, the stereotypes abounded about who and what I was, sometimes regulating me to a position of servitude.

The assumption and subtle messaging in the hiring process that women, and especially Black women, aren't smart enough to succeed in the "hard sciences" like engineering, computer science, physics, or any field of study that requires a high concentration of mathematics, leaves these fields extremely male-dominated. Those of us with the cojones to major in the "tough stuff" find ourselves being "the only—" isolated, unsupported, ignored, and the target of racism/sexism, or both. From as far back as the 1975 AAAS convention of minority women in science, and certainly further back in history, it's proven that Black women have not been emotionally supported and lack mentorship, sponsorship, and critical feedback on work performance, and are ignored when it comes to imperative information required to move to graduate and postdoctoral studies. In fact, "Undergraduate academic counseling was considered poor because of the lack of unbiased advice on course selection, and information about career preparation and advancement, such as the merits of a graduate school, post-doctoral research, fellowships, and internships. Several convention attendees complained that they had never had the benefit of tough, honest academic assessment."[29]

The hard work, long nights, academic rigor, and determination it takes to get accepted into engineering, computer science, or other STEM/STEAM programs pales in comparison to the ingenuity required to complete the program. Sitt and Happel-Parkins, in their study of Black female engineering students, found that "Black women engineering students often find themselves in an uninviting space in a field dominated by White men. Thus, as Black women matriculate toward completion of their engineering degrees, they encounter instances of racism, sexism, and prejudice

---

[29] Shirley Mahaley Malcom, et. al., *The Double Bind: The Price of Being a Minority Woman in Science*, Report no. 76-R-3, American Association for the Advancement of Science, 1976, pp. 17.

that result from the intersection of their race and gender."[30]

If these themes sound eerily familiar, it is because they rear their ugly heads again once we hit the doors of corporate America. In 1994, the United Stated employed only 12% of women in STEM jobs; obviously, a much smaller percentage of that included African American women. This led the Committee on Women in Science and Engineering (CWSE) of the National Research Council to plan a conference, "Women Scientists and Engineers Employed in Industry: Why So Few?" to examine the workplace environment for women pursuing STEM careers. (Are these conferences starting to sound familiar to you as well? Keep reading—there is another coming in another couple of decades.)

One of the major themes that this particular conference uncovered was the stereotyping of male versus female work. We are socialized in American society to think of women's work as supportive in nature, which aligns to housework or administrative duties, and that stereotype carries over into the workplace, where the majority of women entered the modern-day workforce as secretaries, researchers, and other support-based roles. Now layer on the "mammy" stereotype we explored earlier, and the ways in which perception keeps Black women relegated to the bottom rung of the corporate ladder readily emerges.

"It takes the high-profile, high-stakes operational roles in this industry to make Vice President" my new

---

[30] Rashunda L. Stitt and Alison Happel-Perkins, "'Sounds Like Something a White Man Should be Doing': The Shared Experiences of Black Women Engineering Students," *The Journal of Negro Education*, vol. 88, no. 1, 2019, pp. 62-74, doi.org/10.7709/jnegroeducation.88.1.0062.

mentor shared with me on the phone weeks after she joined our company. I was conducting a virtual meet and greet to learn about her background, her new role, and share with her my career aspirations. She went on to say how unfortunate it is that the majority of women in technology companies are in supporting roles, such as project managers, marketing, communications, and HR. "Don't get me wrong, there is nothing wrong with those functions. We need them, and there are so many crazy-smart, powerful women leading those organizations, but you'll never see the Head of HR becoming the firm's CEO. That path is for someone leading a P&L (profit and loss, also known as a cost center)," she said. I remember ending that call thinking, *After twenty years in a support function, where does that leave me?*

The 2020 *State of Black Women in Corporate America* report, mostly compiled by research from Lean In and McKinsey & Co., tells us that "Black women are underrepresented in the workplace for many reasons. One big factor is a 'broken rung' at the first critical step up to manager. For every 100 men promoted to manager, only 58 Black women are promoted, even though Black women ask for promotions at the same rate as men. And for every 100 men hired into manager roles, only 64 Black women are hired. That means there are fewer Black women to promote at every subsequent level, and the representation gap keeps getting wider. If we aren't there to voice our opinions and ideas to ensure minority representation is counted then this means no matter how high we climb, we will be thought about less and less."[31]

---

[31] *The State of Black Women in Corporate America, Lean In,* 2020, Lean In.org/research/state-of-black-women-in-corporate-america/introduction.

According to Chris Weller, "experience bias," as explored by the Neuroscience Leadership Institute, says that the experience we have had with an object or person before will transfer to our experience with a new object or person that reminds us of the previous one. Then, we substantiate that the previous experience was accurate thanks to confirmation bias.[32]

This means that if I am a White male who grew up in a homogenous atmosphere, attended a predominantly White university, pledged a White fraternity, always lived in a White neighborhood, and am now entering a global organization in which I am thrust into working with Blacks, Asians, Latinx, Indian, and other cultures, how I treat them at work directly reflects the views of them I had before I landed in my new work chair. Not having much personal experience I will rely on stereotypes, or my very limited previous experience with them will make it easy for me to judge them, because remember, our brains are lazy and don't like to work too hard. We like to take in new information and process it quickly. This quick categorization of data occurs in seconds and can be harmful if the pattern is not intentionally interrupted. The less diverse the organization, the easier it is for me to treat the African American workers as angry Black women, the mammy, or the Jezebel. Depending on the age of the White man in question, if their previous encounters with Black women included domestic workers such as maids, health care workers, assistants, or nannies, then asking your African American staff member to grab your coffee is no big deal.

---

[32] Chris Weller, "The 5 Biggest Biases That Affect Decision-Making," *Neuroleadership*, April 9, 2019, neuroleadership.com/your-brain-at-work/seeds-model-biases-affect-decision-making/.

If she does, from her vantage point, she is just being nice and thinking this harmless act will get her into the "likability zone." In her mind, it shows that "we are equals" because she plans to ask you to get her a cup next time. However, if she identifies this act as a harmful stereotype playing out that needs to be confronted, you are left without coffee, and she is left in the "angry Black woman" category. Since literature, mentors, and coaches are now advising women of color that we must demonstrate our abilities to manage large teams effectively and efficiently, move into the cost-center management spaces, and not remain in supporting roles, the less and less maid service will be tolerated in our pursuit for parity and reaching the upper echelons of the corporate structure.

I'm a city gal. I've been fortunate enough to live in urban areas with a diverse make-up of ethnicities, races, religious ideologies, and food—can't forget the great food that comes along with different cultures! So, just like the shock of the White man who grew up in an almost all-White environment, I was shocked to learn that there were still people who had practically gone their whole lives without ever having a personal encounter with a non-White person. They only saw us on TV, on social media, and from afar at school. This quickly becomes an interesting, complex, and potentially explosive learning experience, when that same White man, who successfully navigates his way to manager without diversifying his personal or professional social network, now has a diverse work staff to lead. The positive or negative outcome of this new supervisor to lead effectively really boils down to the individual and their willingness to be open-minded and empathetic.

Yes, I, as an African American woman, have all the same considerations of bias and stereotypes as well. However, acting them out and voicing them is unheard of because of the lack of power I hold in society at large and in the company specifically. If I think my new manager is a racist pig, I can't treat him that way. I can certainly be passive-aggressive and push back on his requests or constantly question him to make things harder, but that's about as far as that envelope can be pushed. Furthermore, ignoring him, refusing to complete assignments, asking him to get my coffee in a dismissive tone, would all be grounds for my dismissal —a quick dismissal, and none of the Performance Improvement Plan (PIP) stuff, either. White men are in a power position in the organization as managers. However, more importantly, having power in American society at large presents "the majority" person in this company situation with the opportunity to take in new information, experience someone unlike them or what they are used to, and work to not prejudge or quickly rely on similarity or experience biases, instead doing the work to challenge long-held beliefs.

In her article "Racism and Sexism Combine to Shortchange Working Black Women," Jocelyn Frye also finds that the stereotypes of Black women such as "angry Black woman" and "Jezebel" are damaging to our upward mobility. She reports that negative stereotypes about Black women's attitudes and work ethic assume that Black women do not work hard, resist hard work, must be pushed to perform well, and should be satisfied with any job rather than deserving of the best job.[33]

---

[33] Jocelyn Frye, "Racism and Sexism Combine to Shortchange Working Black Women," *American Progress*, Aug. 22, 2019, americanprogress.org/issues/women/news/2019/08/22/473775/racism-sexism-combine-shortchange-working-black-women/.

These stereotypes hurt our career progressions and abilities to be perceived as intelligent, capable leaders while also making it extremely hard to focus on the work versus managing others' impressions of us. The compulsory mental hoops to assert myself as a leader while delicately balancing the tightrope of remaining likable is near impossible.

The most critical issue to understand is that as long as African American women in scientific fields continue to serve as the meeting minute-takers, the coffee-gophers, the presentation-creators, and the behind-the-scenes workers, we will never get the opportunity to shine. Without the chance to present to senior leaders, lead a team, and ascend to eventually run a cost center, we will never have a chance to knock on the door of the C-Suite, let alone enter it. Unfortunately, we learn from Lean In's 2019 research that even when we do get those opportunities and "kill it," the positive outcomes are attributed to external factors and not our innate abilities to lead.[34] In order to dismantle the notion that our positive performances are one-off flukes or strokes of good luck, Black women need to highlight that we indeed can contribute consistently and effectively. We need the opportunity to do so and nail it repeatedly, thereby making it increasingly difficult to dismiss our work.

It makes good business sense to enable those with the passion and capability to lead, but there is an intangible benefit to the workforce as well. Despite the perceived meritocracy in corporate America, the "old boys' network" is alive and well in many firms, and it is thriving. So, if White women and other minorities see

---

[34] *Women in the Workplace 2019*, Lean In and McKinsey & Company, Oct. 2019, womenintheworkplace.com/2019.

African American women succeeding, that sends a very clear message that all are welcome and included.

Anjuan Simmons, a Diversity, Equity, and Inclusion (DEI) consultant, employs the Althea Test to help organizations with their DEI initiatives. She asks the C-Suite team three questions, and only three, to quickly assess the maturity of the firm. They are:

1. If there are Black women who are individual contributors in your organization, do you know their names?

2. Can you list at least two contributions they made in the last quarter?

3. Do you see them doing your job eventually?[35]

An affirmative response to all three inquiries indicates a highly inclusive environment, and a negative response to any of the three means the company has quite a lot of work to do. Simmons argues that "due to intersectionality (a term coined by Professor Kimberlé Crenshaw in 1989), Black women are often doubly penalized by race and gender discrimination. Furthermore, initiatives aimed to combat gender discrimination often miss Black women because of their race while those designed to combat racial discrimination overlook them because of their gender . . . Providing assistance for the most vulnerable provides a much more effective system of help for everyone."[36]

---

[35] Anjuan Simmons, *The Althea Test for Measuring Inclusion Maturity*, anjuansimmons.com/blog/the-althea-test-for-measuring-inclusion-maturity/.

[36] Ibid.

Maid Service

The subtle ask for help or support from your team seems innocuous at first. There should be no harm in taking meeting minutes or grabbing coffee for a coworker—that is a sign of being a team player. However, it becomes maid service when it is expected of you to do it. The shift can occur slowly or quickly, but when it happens, it becomes a hindrance to your career development.

I had just joined an organization a month prior, and as the new kid on the block, it was my turn to organize the team's Halloween decorating contest. I dove in with both feet and hands! Never one to decorate for Halloween, though I'd won a couple of costume contests back in the day, I was game to put my project management skills to the test for a friendly floor-to-floor adornment competition.

I had fun researching theme ideas, meeting new team members to brainstorm, assigning tasks, and found that making trips to the store to purchase materials and decorations was a well-received break from the stress of my former job. It was a success—our department won, and guess what? I got to do it again the following year. And guess what? We won again that year too! It was fun to orchestrate the Halloween decorating contests, as well as planning the subsequent team-building activities at the local bowling alley and contracting with trainers to come speak to the team about developing themselves to work together more effectively and improve our cross-team communication. Honestly, the fun field trips were a nice departure from challenging work and the competing priorities we experienced. However, I quickly realized that it was always falling on to me to organize, and by the end of year two, it had become an expectation.

By this time, I had now connected with mentors and advisers and attended senior management networking and professional development sessions that were all singing the same tune around impression management and how critical it was for me as a woman to be seen in more hard-hitting roles and delivering bigger and bigger initiatives. I wanted to shift my image. I needed to quickly shed the "fun girl" persona and portray my leadership skills. I declared that I wouldn't be managing that year's big team-builder and we should look to another manager to orchestrate Halloween. Unfortunately for the team, that was the end of "the fun stuff" since all the other male managers didn't take up the mantle, and the only other two women managers had a lot going for them, but "fun" wouldn't have been a word used to describe their personalities or interests.

    Another moment of my apparent maid service that stands out in my mind is when I was told, "You just need to do what I tell you to do." That statement stung and stung hard. I had just returned from maternity leave temporarily, as I already knew I was going to be looking for a new job based on what my colleagues had been telling me about how the new CIO had come in and completely changed the culture of the IT department. Now, let's be clear—it wasn't a stellar culture by anyone's definition. The company would have failed the Althea Test and any other DEI assessment before, but the new CIO stamped out any hope of progress, especially for women. I saw him ignore my White female supervisor and cancel one-on-one meetings with her to the point where they hadn't directly spoken in weeks. So, if the workplace culture was suppressing her, you can imagine that the weight of oppression on me was suffocating.

    Apparently, a week prior to my return to work, a new Head of Architecture was hired, and his first day would be during a major planning session. Basically, it was a large multi-department meeting in which all the leaders come together to share the vision, mission, and goals, then create a roadmap for the work ahead. Our sessions were usually twice a year, but we were moving to a cadence to increase them. Upon my arrival, I was dealt the card of the session leader. In this first session, the vendor was included, which was rarely done, but in this case, it made sense as they were an integral part of helping us develop our strategic roadmap. A disagreement between the new head of architecture and the vendor blossomed into a small war as they butted heads over which functions the system upgrade could deliver and if subsequent integration would be able to provide the requirements communicated by our business partners.

In my opinion, logic won. Our vendor representative not only stated that the functions we wanted would not be available, but we also had to arrange a call with the head of product development for the system who clearly stated that the system would not perform the way we'd expected. Therefore, it seemed like a rational action to revise the plan to perform the system upgrade and integrate it with a different solution to achieve the outcomes we needed for our customers. I quickly pulled together the plan proposal, vetted it with the subject matter expert, and prepared my pitch to the IT Director.

The Head of Architecture sat pacing in the IT Director's office and fidgeting while I spoke. He tried to interrupt me a few times, but the director was professional enough to snap back at him, saying, "Let her finish." When I had finished presenting the plan, the director made the usual cop-out move, which was him saying, "I will let you both work this out." Which was code for, "I don't want to be responsible for making a costly decision here and put all my eggs in either basket, so although the little Black girl makes sense, I'm not making a commitment either way."

This led to me begrudgingly having a private meeting with the "jerk of architecture" the next morning. I was prepared to negotiate and meet him more than halfway, but instead, the moment we both sat down, he looked me in the eye, chest puffed with White male dominance, and proclaimed, "You just do what I tell you to do." Incredulously taken aback, I quickly threw up my filter so expletives wouldn't fly out, like, "You sexist asshole!" Instead, I took the "we are on the same side, wanting the best outcome for our customer" stance. He slowly blinked and said, "Look, you are the project manager. You just put a schedule in place that tells the team what I want them to do." After a comment or two I realized that he was a brick wall and had made up his mind that I was dirt on his shoes and good for little more than making copies and setting up meetings. I documented my objections, created the schedule he insisted upon, calmed the team members' fears that we were running a fool's errand, and sat up more interviews to get out of there before that project crashed and burned and I was blamed for it.

Fast forward to a little over a year later, I was gone, and the other Program Manager I'd turned that project over to when I departed had joined me at my new company. During a lunch session, I was afforded the opportunity to learn what had happened after my "sudden" exit. The architect had insisted that he was right and the use of some middleware between systems would get the customers the functionality they desired, and he tasked my successor with managing it (after the right contractors were found, of course) since he thought our IT staff was too stupid to do it. After a year of contracting a boatload of high-expense contractors, procuring the hardware, and cobbling together a system

Maid Service

that barely functioned, the head architecture jerk finally ticked off the wrong person and was let go.

I was mostly relieved for the staff as they were finally free from the tyrant dictating orders and blasting their projects out of the water. I was relieved for myself because I knew, as the only African American female Senior Program Manager with zero support, zero mentors, sponsors, or advisors in the organization, when that project had gone belly-up and turned out to be less than successful, I would have been blamed. If the project were successful, the architecture jerk would have taken the credit. Turning it over to my peer, who was White, male, and an extremely decent guy and solid Program Manager, I knew he would survive had things gone south, and if it had gone well, he would have been on top. In summary, a win for him either way, and I got out of being the help.

Being talked down to, belittled, and treated like the help gets me all fired up and hot under the collar. Being ignored and treated unintelligently in front of my peers is embarrassing and humiliating. One time, about 2004, I was pretty much hoodwinked into being a project manager (PM) when my department reorganized, and I lost my enjoyable business analyst role. I was told that moving into Judith's department as a PM would get me the mentorship and coaching that I needed to progress in the company; after all, Judith was well-known for getting women promoted. So, off I went to Judith's team, where Judith announced her retirement two months later. Left with no supervisor, I had to fend for myself after my assigned "project manager mentor" had plopped the Project Management Book of Knowledge (PMBOK) on my desk and said, "There, follow that."

After three failed nights of reading the book, which made for amazing sleep material, I was slowly starting to give up. Thanks to a benevolent and empathetic manager, who we will call "Brandon," who reached out and showing me the ropes, I picked up on the tools and processes, enough to be a decent project manager. By the end of the year, they announced a new manager returning from the field who would lead our team.

Day One, we started off on the wrong foot. He walked into the room of all women (we were Judith's legacy of assembling an all-female team, except for Brandon), and the new manager nervously laughed and said, "Now I know how Angel and Lynda feel."

Lynda and I were the only two African American women on the team out of the entire 120-person Stats and Data Department. With that first comment, my bias was triggered, the filter went up, and my side eye initiated! And I was right—after working directly for him for over two years, I can without hesitation proclaim that he ascribed to the "all Black folks are lazy, can't be trusted, and need an overseer" mentality.

The experience for Lynda and I was rough, to put it mildly—rougher than the other White women on the team had it. By the time my third project kicked off, I was slowly starting to build my confidence, and with Brandon's help, I was stepping into my own. I knew I was making advancements when technical subject-matter experts were proactively asking me questions about the project and sharing status updates. Additionally, the new Stats and Data IT manager, "Kevin," partnered with me to manage the project and we worked very well together.

During our weekly project execution review session, I reported that the project was officially in the red since we'd added scope and lost two full-time developers to a higher-priority project.

My supervisor turned to Kevin and asked, "Why is the project running over?"

Kevin said, "Well, just as Angel stated, we added scope and lost two full-time developers."

My supervisor nodded his head and said, "Oh, okay." After the hundredth time of this happening—okay, I may be exaggerating, but it did occur weekly, and every week, Kevin would become increasingly peeved that he would have to parrot my report. We often exchanged incredulous looks, which I was appreciative of, since it offered an unspoken "you aren't crazy" message. In case you are wondering whether Lynda had suffered the same fate, the answer is yes. For any projects she managed for Kevin, he would have to repeat her status updates as well.

One fateful day, I had simply had enough. When my supervisor asked Kevin for the report after I had already provided it, and Kevin quickly whispered the update, I said, "Is there an echo in the room?" The whole team snickered, and my supervisor looked up from his notes and smiled, unaware that the joke had been at his expense. When the meeting concluded, I marched up to him, towering over him in my best trying-not-to-be-the-angry-Black-girl but firm and assertive stance and tone to inquire, "Why do you ask Kevin for the report after I've already given it?"

He looked up from his chair like a little boy being scolded for getting dirty while playing in the backyard in his church clothes. Then his face shifted to a blank stare, and he looked back down at his notes. He started shuffling them, fidgeting and murmuring with no clear response. I pivoted on my heels and marched out the room to Kevin's desk. I asked him, "Kevin, do you notice that you have to repeat the status report after I do?" He looked sad, whispered, "Yeah," and shrugged his shoulders. I told him that he could help stop it. He nodded his head declaring, "Yeah, I guess I can."

After that, Kevin would defer to my initial report until one day it got contentious, and my supervisor kept pressing Kevin to give a detailed report. Kevin had finally reached his wall and asked, "Haven't you noticed I just repeat what Angel says?" My supervisor blinked, and we all left in awkward silence. Kevin's pushback renewed my courage once again, so in a one-on-one session with my supervisor, I asked him, "Do you want a different format? Do you need more detail?"

He stared at me blankly and retorted, "I just need to hear it from him." Then he stood up and left the room.

Frontline Tales

> Every meeting began with me acting as secretary and taking the group's meeting minutes. One day, I'd had enough and said, "I'll take them today, but who is doing it for the next meeting?" Blank stares and lots of blinking ensued. After a long pregnant silence, Jimmy volunteered, and the meeting started.
>
> —Helen, IT Director

Being mistaken for the maid, sometimes literally, can happen no matter how high up you are in the corporate hierarchy. At a summer IT conference a Black woman CISO (Chief Information Security Officer) recalled her story of being asked to get a table mate's coffee among a table full of people.

Maid Service

> A middle-aged, White male conference attendee, one of many in a sea of White faces (as I was the only chocolate chip in attendance at this conference—not surprising, by the way) at a Cybersecurity Conference asked me to get his coffee. Long story short, this became a teachable moment, because I did get his coffee. I sat down next to him and proceeded to chat with our table mates. The MC walked to the stage, and after a few introductory remarks, they introduced me, the keynote speaker. I slowly turned to my left with the widest, toothiest grin I've ever brandished, nodded at his completely red, wide-eyed, slack-jawed face, and I proceeded to the podium. Fast forward—we are friends to this day, and I am his mentor.
>
> —Janice, Cybersecurity CISCO

Oftentimes, being relegated to the maid is very much unconscious bias playing out. When you see it happening and recognize it, calling it out can help reposition your standing at work as a capable and intelligent worker who can complete challenging assignments and not always take notes or fetch drinks. After all, none of us sacrificed (and many times, none of us had our families sacrifice) everything it took to graduate from college to not use the degree that was conferred to us. In short, whenever they try to "put baby in a corner," you can resist.

> I could process a lot of information quickly and kind of spit it back out as a BA [Business Analyst]. I was up for a promotion at the time as a manager, and I got passed up for promotion the first year. And I remember I was managing

people. I was doing everything that I should have been doing, but I didn't get the promotion because they said I wasn't known for anything. So, I was like, "Okay, fine. I'm about to be known for something." And I was saying, "Okay, well, I'm going to go be an agile coach, because this is hot right now." I'm like, "Okay, well, I'm about to go be an agile coach." Mind you, he was, like, three levels above me. But I was like, "Okay, well, you said I'm not known for something. You said I'm not up for promotion. So, I'm going to go be an agile coach, just like you're an agile coach." Right?

. . .So, then I became like the auntie, the mama, whatever, of the group, right? Because it was these older executive men... and they needed somebody to put together the decks, to do the research, and to put together our point of view on that topic. And I was that somebody, right? My managers would treat me like an intern. I was getting coffee, and I'm like, "Why am I getting coffee? I'm a consultant."

—Vivian, Agile Coach

And so, for us, we were rolling out a series of educational products. We have a four-year apprenticeship program, and I did revisions to the curriculum. "She can't do that. That's outside of her scope. She's an operations manager." And so, I'm like, "What makes you think that I can't update the curriculum? What about my title has anything to do with the copy-editing skills, the research skills that I have?" And so I had to continue with the opposition and actually prove to them through my version of changes versus the changes of someone before me to say that I can actually do the job. . . . I had to actually prove it to them, and I had to do it a couple of times before they stopped second-guessing me on my capacity.

So, I'm good enough to schedule the meetings. I'm good enough to facilitate. I'm good enough to make the changes. But I'm not good enough to be the editor on it. I'm just not good enough to write my name on it, and that was the thing. And they wanted someone else to, "Well, we're going to get someone else to review it, and then they'll put their name on it." And so, I'm like, "That's not going to happen. So, tell me what they've done." And so, ultimately, I got my way, but it took some time for me to get them to understand what's fair and what's transparent and that it's kind of discrimination.

—Carol, Chief Operations Officer at Electrical Engineering Company

...When I left, there was one other Black female director who had been there for years and should have been a VP. She needed to leave, right? She ran this whole organization. Again, the cleanup woman.

—Paulette, Senior Agile Coach

That's probably one of my biggest strengths, but at the same time, it's also probably one of the biggest detriments to my career trajectory because when you are the SME, or the subject-matter expert, in certain areas, and people are naturally—maybe not even naturally, but they're just prone to come to you for support—then that burdens you. . . but for sure, for me, I'm mostly the person that's taking the minutes, doing the reporting, putting together the PowerPoint decks, and presenting the information.

—Jalessa, Process Manager at Financial Services Company

Okay, I've got the engineering part. I'm doing the engineering role. I'm doing the business part. I'm writing the contracts. I'm doing the legal part. Now you all have me doing logistics? One-person company over here."

—Lisa, Cybersecurity Analyst

And sometimes people thought I was going to be the person taking notes, and I was like, "I'm not doing that, because I'm the subject-matter expert. I'm the lead. I'm not going to do that." And I've had to assert myself that someone else can take notes or do things like that in instances. And a lot of times, people—I mean, it's like, "Well, who does she think she is?" And in those instances, you have to decide.

—Reagan, Senior Consultant at Big Five Consulting Firm

I think there were instances where this same particular manager, I think, more so when we were in different roles—because I had been under this person several times in my tenure at that company—would say certain things to me like, "Well, can you go schedule that meeting?" It's your meeting. And so, we'll be in the meeting, and she would be the one initiating and agreeing that we should have a meeting, but then turn around to me and say, "Well, can you handle scheduling that for me?" Excuse me. That's not in my job description. This is your meeting. Like I was her secretary. I'm like, "Your calendar shows you the same conference rooms as mine."

—Kamryn, Business Analyst

"Oh, you can organize a team to get, maybe, our monthly report out." That's not my forte. That's not my passion. I'm not good at that. . . .Maybe because I'm on time with my assignments means that I will like or find value in organizing the rest of the team to get to their assignments.

—Tracey, Director of IT at Electric Manufacturing

I became "hun" and "sweetie." "Go get my coffee." And I was like, "I have a degree from Purdue. You're going to get your own damn coffee." And I'm not a quiet person.[I maintain my stance of not performing admin duties]

[Then later, at a subsequent job,] when they were like, "We need somebody to take notes," I'm looking around like everybody else is looking around. "Nope. Not it."

—Kimberly, Web Administer of Credit Services Agency

Dents in the Ceiling 151

You can learn to shift out of the servant role gradually and under the radar, or boldly resist, like some of the ladies we just heard from. Your approach will be influenced by your department or team's culture, your personality, and your comfort level. Properly assessing these three aspects of your unique situation will inform you as to which clue to use to resign from maid service.

Maid Service Clues

Kindergarten Rules

What is the fundamental rule that we all learned in kindergarten? Yes, that's right—the rule of sharing. In addition to writing your name, tying your shoelaces, and climbing the monkey bars, you learned the golden rule of equal distribution. It quite simply means to take a turn. Only one yellow crayon is left, and you really want a pretty, shiny, golden sun in the corner of your picture, but your best friend has it? Time to throw a fit and snatch it from their pudgy little fist? "No," says the teacher! It's time to learn how to take turns.

The concept of turn-taking, along with basic politeness, has been lost along the way in corporate America. Apparently, the higher your climb on the career ladder, the less and less sharing needs to happen. Old-school command and control rules took over in industrial business and have cemented themselves in the working relationships of the tech industry as well. When those in power assume that you will be the one taking minutes, now (meaning your very next meeting) is a great time to remind them of the basic principle of sharing. There are a few ways to institute this forgotten concept for your team.

Start the meeting by saying, "I've created a new meeting-minute template. *Jill*, do you mind if we collaborate after the meeting and I'll show you? Perhaps you can use it for our next meeting?" You just called Jill out very politely. She may try to back out by saying, "Oh, you do such a good job, and I'm really bad at it," but don't take no for an answer. Respond back by commenting how easy it is and that she can just follow your format. You can offer to show the next person after Jill, so Jill knows it is just one time, and the rest of the team is now on notice that they will be called on, and minute-taking becomes a team sport.

Another tactic that is a bit less confrontational is to ask an ally or the least-combative person on your team ahead of time to take the minutes before the meeting even starts. You'll have to judge the timing of the ask. Don't ask too soon, which gives time for "something to come up," but you don't want to spring it on them at the last minute, especially if you are trying to be kind and not do a "call out" as in the first option.

Finally, another technique is to be bold in your stance. You can announce your intention of making minute-taking an activity that is to be shared at the start of the meeting. Such as, "Before we get started, I'll take the minutes today. Who would like to volunteer for next week's meeting?" Now, sit there and allow the awkward silence to settle. At least one other slightly emotionally intelligent person will capitulate, handing you a promotion out of indentured servitude.

Resist the Force

As you learned from a few of the ladies I interviewed, they never went along with forced expectations from the start. As my grandma would say, "You teach people how to treat you." If you never succumb to other people's expectations to be the leader of the "fun committee," the meeting-taker, the coffee-fetcher, or any other act of service to the team or organization, then you don't have to worry about how you will shed that image later.

Learn from other African American women who have come before you and set the expectation from Day One that you are not the one. This means that you don't volunteer for those duties unless you believe that they will serve you and your career in a positive manner. I caution you to watch the trap, as you have enough work to do with impression-management and being a capable, smart, and valuable leader in the organization. No matter how great your brownies or red velvet cheesecake is, keep those recipes to yourself.

If baking is a strategic advantage on your team, then by all means, get your Betty Crocker on; otherwise, think twice about which activities you choose to engage in that are outside of your core work duties. You will have to play outside of the nine to five in order to move into the "likability zone," but it should match with what those in leadership are doing, which stereotypically is golfing, boating, and attending sporting events, music concerts, or art gallery openings. This is your opportunity to leverage your cultural competence once again and engage in outside-work activities that your organization's leadership finds valuable. The stress that comes with the expectation of baking brownies for every birthday, baby shower, or promotion can be overwhelming depending on your team size. It may not build you any political capital, but organizing a one-time department-wide community service project may just be the ticket to add a few more tokens in your political capital coffers.

There is nothing wrong with saying "no" in a polite way. You don't have to accept everything that is offered to you. "Tameka, would you like to lead our Christmas party decorating contest?" There are many ways to approach this question, and depending on your organization culture, some responses are more appropriate than others. I invite you to think about alternatives to achieve what you want. For instance, if you are asked in a private setting, depending on your trust level with the person asking, you could be honest.

First, ask them why they asked you. Was it for your keen organizational skills? Did they see a picture of your house decorated on your screensaver? Did someone else recommend you for your creative abilities? You are seeking the thought process and logic behind the request. If there is no objective rationale for asking you, was unconscious bias at play? Did they just assume you would want to do the work, and if you're not interested, no big deal? Do you sense that if you decline the request, there will be negative repercussions? Is this a "volun-told" situation? The important point is not to get stuck, so if you feel that you have no option but to get your jingle bell on, then solicit help. Secure a second-in-command and set the expectation that you plan to lead this time, and the next time, it will be their turn.

This will be an opportunity to showcase your delegation skills, your ability to work well with someone else and train up a junior person (even if you yourself are a junior person—in this situation, this person is junior to you). Did winning the Christmas decorating contest under your tutelage not earn you any stripes? Reframe the work you completed on your performance review in functional terminology that includes words such as delegation, strategic planning, and cross-functional coordination.

Outside Work

This is my favorite clue of all time! I mastered this one, which not only got me out of all that mundane maid work but landed me a promotion and propelled my career forward. The secret to my success is what we will call "outside work."

Outside work consists of all your professional organization positions, conference-planning assignments, side-teaching jobs, community service work, and even side-business hustle. Doubling down in these areas will help you grow your expertise. We hear stories all the time of managers not listening to their employees, but an outside consultant comes in and gets paid thousands of dollars to tell leaders the exact thing employees have been saying for years. There is something about an outside source that lends more credibility. So, my dear, it is time for you to become that outside, credible source to someone else.

Start simple by joining a Meetup. Meetups are usually free and an excellent way to network. Most Meetups are virtual now, so the barrier to entry is low and time commitment is minimal. After contributing to the Meetup group, you can volunteer to share your expertise.

For example, an attorney once gave a talk to a project management Meetup group on basic contracts to help PMs learn more about the legalities of statements of work for the projects we deliver. His original purpose was attending the Meetup to get tips on getting more organized and using online scheduling software. His talk was a huge hit, and he propelled himself into expert status to a group outside of his law firm. The law firm may not have been giving that attorney big-time, career-making cases, but once it gets out that this attorney is speaking to other groups as an expert, I'm sure his firm will notice his new reputation.

Another action you can take is to join conference-planning teams. Planning an in-person or virtual conference will allow you to see how the "sausage is made." If you want to be a speaker at a conference, develop a solid understanding of how speakers are selected, which conferences pay for speakers, and the logistics of speaking; all of this can be learned after one or two conference-planning cycles. You can lead various committees in order to learn all aspects of the planning process.

Leverage this knowledge to be a friend to other conference planners when you offer to speak to their audiences. Be sure to pitch a topic you can speak on in your sleep, as this will build your confidence. And let's face it—these conference planners really need to increase their speaker diversity anyway! Take the technology and processes you've learned on your job and share them with a community that needs your expertise.

For example, I eat, sleep, and breathe Agile. I practice Agile principles, processes, tools, and techniques every day at work and at home. I love sharing how an Agile Mindset can help anyone, from someone trying to lose weight to someone starting a new business. I'm also a Black, female Agilest, where the experts and formal coaches are all a sea of White men. I can talk to any audience about Agile, real project management (not the textbook way), and DEI of tech. What are you an expert in that could be of help to an outside group?

Finally, have you thought of going back to school? Not as a student, but this time, as a teacher? Community colleges love industry-knowledge adjunct faculty. We aren't textbook and theory; we have lived through the late-night cutovers, the server failures, the missed project delivery, and the security breaches. We have been there and done that and have a great deal of real-world knowledge to share with students.

I have personally applied all of these outside-work activities. Not only did I get a healthy self-esteem boost, which was desperately needed after being told I wasn't technical enough or that I was too passionate or fill in the blank, but I was also starting to become an expert outside of my organization, to the point where the company couldn't ignore my contributions to the tech community at large any longer. Leveraging all these outside-work initiatives expanded my network and put me in direct contact with influential leaders, not just locally, but globally.

Your final clue to stop being the maid and a leader instead is your social media presence. Please go take a professional marketing class. It was one Saturday and less than a $100 investment for me to learn how to self-promote on social media. Phenomenal head shots and a slick LinkedIn profile are both stable stakes for any serious tech professional. If you aren't being heard within your company, posting technical research you've done on your own time, sharing your expertise, or blogging about an industry-leading topic are all excellent ways to perform a rebranding for yourself. Just because you've been typecast as the maid doesn't mean you can't work yourself into a new part.

Active Ally Clues for The Help

Boy Scouts

Even if you were never a Boy Scout as a child, it's never too late. The Boy Scout slogan is "do a good turn daily." Consciously seeking opportunities to help someone else is your job as an active ally. To accomplish this, you do not have to hold a position of supervisor, manager, director, or above—just a willingness to see how African American women on your team are being treated. If you don't like what you see, you can work to change it.

Keep in mind, you don't have to play the role of savior. "Friend" will do just fine. Leveraging your active listening skills from earlier to understand a Black female colleague's biggest challenges at work is important. If her issue is being treated like the maid, and she is always being assigned to supportive activities, you can bring attention to that fact by volunteering to perform those duties, work alongside her, or just offering to brainstorm ideas to help her with her rebranding efforts.

Below, Karla, a fellow in tech, was pregnant when a coworker launched a donation collection for her baby shower. A barrage of emails came in, in which she was included, about how people did not want to donate, asked to be removed from the list, and that they didn't even really know her.

> So, [some] people were like, "Don't take it personally." And one of the guys on my team who I never really felt close to, a younger White guy, actually replied. He's like, "Look. If [they] don't want to participate, that's totally fine." And I never felt really close to him. We had kind of gone to lunch a couple of times when the whole team went.
> —Karla, Information Security Manager

It doesn't take much effort. In Karla's situation above, her colleague reaching out to talk to her after a negative interaction with her colleagues was helpful.

She was surprised that someone who she wasn't that close to recognized her unfair and embarrassing treatment and commented on it. Offering a listening ear and working to actively support someone may not earn you a Boy Scout badge but definitely an Active Ally stripe!

Girls Scouts

Women helping each other in the workplace is not a given, especially when race enters the picture. White women have started to admit the power they have in the workplace and are slowly, VERY slowly, giving up their victim statuses. White women have benefited from diversity and inclusion initiatives since their inception. Tech companies have often flown their diverse flag high once White women entered the executive suite, since gender is the more acceptable diversity characteristic for White men in power. Ultimately, White women stepping into leadership roles for tech have been more accepted than any other diverse population.

Megan Bigelow noted in an article for *CIO* that "it is incumbent on the White women in the 'women in tech' movement to course correct, because people who occupy less than 1% of executive positions cannot be expected to change the direction of the ship. . . .It is incumbent on White women to recognize when they have a seat at the table (even if they are the only woman at the table) and use it to make change. We need to stop praising one another—and of course, White men—for taking small steps towards a journey of "wokeness;" instead we need to push one another to do more. At the very least, we must acknowledge these truths and give them visibility."[37]

Reaching out to listen, speaking up to encourage an equity share of supportive tasks for the team, and volunteering to share in the workload are all actions that White female allies can take. The division between White and Black women at work has been long and vast and spelled out in the book *Our Separate Ways: Black and White Women and the Struggle for Professional Identity* by Ella Bell Smith and Stella Nkomo.

It will take a while for us to reach the point of truly supporting each other, but as for White women working in tech, you know how hard it is navigating the male patriarchy. You've learned some things along the way. Instead of looking to the men to make the workplace more inclusive, you have power to do so as well. White female allies, you don't have to take an oath like the Girl Scouts, but to be an active ally, you should. Maybe it is time to pledge, "On my honor, I will try to be an active ally to my fellow sisters at work and do everything within my power and authority to make my team and department a place of respect, equity, and belonging."

Bust Barriers

Active allies help break down barriers. There are informal behaviors that directly impact the team, the department, and the overall culture of the company. These informal behaviors ultimately drive outcomes for the company, and these outcomes can be positive or negative.

---

[37] Megan Bigelow, "Ending white dominance in tech starts with white women," *CIO*, Nov. 1, 2019, cio.com/article/3451516/ending-white-dominance-in-tech-starts-with-white-women.html.

It costs a great deal of money to attract, recruit, and on-board diverse new talent while retention is almost non-existent. Why hire a 4.0 Information Science grad with an MBA and three years of experience to treat her like a secretary? This just doesn't make good business sense. Ignoring your diverse population and relegating them to the role of the help is a sure-fire way to kill creativity and stifle innovation, and that is if they choose to stay.

Unleashing the power of your team can start with small steps. Think about what stops people from working together, what sucks their time, and what they need to do for their jobs to be more effective. Then start breaking those impediments. You don't have to be in a place of management to think like this; in fact, it might help your personal career if you are an individual contributor.

One of my new Project Coordinators was handling all the paperwork for the PMs on the team, such as processing vendor timesheets. Manual timesheet submissions seemed archaic to the contractors, too, but we were doing it old school. She declared that she would have more time to support project management activities that mattered, such as managing risks and issues and updating critical status reports, by automating the timesheet submission work. It wasn't easy, but she gathered the requirements, advocated for the system upgrades, and successfully removed the admin work from her plate. Most people reading this would think, "Wait. . .she was a project coordinator. Isn't the majority of her job administrative work?" Yes, but if she had stayed stuck doing all admin work, how would she ever have had time to demonstrate her ability to be a project manager?

This is the exact conundrum in which early-career to mid-career diverse talent finds themselves often. If they aren't freed to be more strategic and less tactical, how will they showcase their abilities? Therefore, they are forced to perform outside work or must push back on support-type work because they are justifiably fearful of getting stuck. Now is the time to think critically about these barriers to success for our diverse team members and know that removing them will benefit everyone.

Breakthrough Tools

- Being overlooked or ignored at work is obviously detrimental to one's career trajectory.

- Taking meeting notes, setting up before an event, or getting a coworker a drink during break are all "team player" activities; it is when you are always the assumed person to perform these tasks is when it is time to take notice.

- Create a strategic plan that will showcase your impact in ways that are important to your supervisor and senior leadership.

- When asked to execute a support task, get clarity on why YOU where asked, set clear boundaries of how often you will perform the task(s), and master how to communicate the result of what you did to help reposition you from "maid" to "impact driver."

Starting Points

Women: When caught in a cycle of performing support work, enlist a junior team member, or someone else on the team, to transition the work to next.

Allies: Volunteer at the next meeting to take the notes. Too senior or too busy? Specifically ask another team member to share the load of the administrative work for the team, then follow-up to ensure that they helped.

Chapter 5

Emotional Misconduct

"If I can't work with you, I will work around you!"

—Annie J. Easley
Computer Scientist/Mathematician/Rocket Scientist

African American women in corporate America are at the bottom of the totem pole. We are the lowest rung on the hierarchical ladder when it comes to the socio-political climate of the company. This low-ranking position makes us vulnerable. This leads to exposure to not only discrimination but sexual harassment. My personal experience, along with the experiences of many of the women who spoke about their work-related sexual provocations, focused on the suffering of the emotional exploitation that left wounds of doubt and mistrust that are still trying to heal to this very day.

Ranking of people does not start within the organization itself. Corporate America is a microcosm of American society at large. The pecking order of heterosexual, able-bodied White men remains at the top of the pyramid, while African American women remain at the bottom and other minority and marginalized groups cling to various rungs in between. This structure is set the moment a company opens its doors, and any business that shakes up this positioning has to work exceptionally hard to do so.

In the book *Caste*, Isabel Wilkerson defines how being a part of societal assignment occurs. She writes that ". . .caste is the granting or withholding of respect, status, honor, attention, privileges, resources, benefit of the doubt, and human kindness to someone on the basis of their perceived rank or standing in the hierarchy."[38] Racism and casteism do overlap, as she notes that "what some people call racism could be seen as merely one manifestation of the degree to which we have internalized the larger American caste system."[39] Casting occurs at birth, and one must sacrifice a great deal to propel themselves into a higher order.

Our casting as Black women exposes us as it puts us at risk of being preyed upon with very little recourse to speak out. Black women's experiences with racism and sexism are complicated by their position at the intersection of racial and gender oppression, which is sometimes termed "gendered racism."[40] If you are "the only" and *already* have issues with being heard in a

---

[38] Isabel Wilkerson, *Caste: The Origins of Our Discontents*, Penguin Random House, 2020 pp. 70.

[39] Ibid.

[40] Anita Jones Thomas, et al., "Toward the Development of the Stereotypic Roles for Black Women Scale," *Journal of Black Psychology*, vol. 30, no. 3, 2004, pp. 426-444.

meeting, you feel excluded from social events and struggle to show your value for a promotion, then the likelihood of a positive outcome if you report being sexually propositioned is extremely slim. This usually leads to one of two conclusions: stay silent or leave.

Sexual harassment of African American women in the workplace is perpetrated by both White and Black men. Black men—regardless of status—can feel comfortable making sexual advances toward their "sistahs." These men are often "the only" as well. African American men with successful STEM careers have often achieved a socio-economic status that is above that of their friends from high school or college. This higher income affords them the opportunity to live in upper-middle-class neighbors and operate in upper-echelon professional social circles that do not include other African Americans.

The opportunity to work with another "only" avails itself to the excitement of being able to be one's authentic self. This means dropping the code-switching, or the "mask," as well as the ability to connect over cultural music, local social events, and the unique perspective of being Black in corporate America that only they can share. This cherished and potential supportive connection can all too quickly turn sour with one flirtatious hand on the knee or inappropriate invitation back to a hotel room during an out-of-town work trip.

African American women who already wrestle with impression management now must deal with the lack of emotional support and one less ally of the other "only." It is a tremendous loss. Anita Hill's accusations against Clarence Thomas is a famous example of how African American women who call out the unprofessional, inappropriate, and sometimes illegal harassment committed by their "brothas" can be met with contempt, even from their own community. It was demoralizing to see how this particular confrontation reinforced the perception that any woman who raises the issue of sexual oppression in the Black community is somehow a traitor to the race, which translates into being a traitor to Black men. It is particularly disheartening knowing that a lot of Black people took this stance despite believing Anita Hill. They who decided that standing behind a Black man, even one with utter contempt for the struggles of African Americans, is more important than supporting a Black woman's right not to be abused.[41]

Judy T. Ellis reported that when respondents who filed sexual harassment claims at work were asked if they were filing racial or sexual harassment claims responded that they could not tell.[42] Although sexual harassment is about power and control, the intersectionality of Black women's gender and race are so intricately tied that no one can really distinguish why we are being harassed. Is the notion of the Jezebel stereotype playing out, or is it because the aggressor believes that all women should be at their disposal? At the end of day, although the origin of provocation is an interesting psychological study, who the object of the persecution is makes no difference. The only answer is that it should stop.

Instances of the sexual harassment of Black women in the workplace is where the outside culture of America at large begins to be mirrored within the

---

[41] Marcia A. Gillespie, "We Speak in Tongues," *Ms.*, Jan.-Feb. 1992, pp. 41-42.

[42] Judy T. Ellis, "Sexual Harassment and Race: A Legal Analysis of Discrimination," *Notre Dame Journal of Legislation*, vol. 8, no. 1, 1981, pp. 41-42.

corporate climate. American society has a "blame the victim" mentality, especially if that victim has very little political capital. Women are blamed for their victimization as soon as anyone tries to generalize, contextualize, or explain away their abuse by saying:

> "You asked for it!"
>
> "You put yourself in that situation."
>
> "Why didn't you come forward sooner?"
>
> "Why did you let the abuse last so long before speaking up?"

These are the questions any woman—regardless of color—faces if the reports of her abuse go to the police or even a company HR representative. To avoid this second blow of hurt and victimization by being made to feel like they are at fault, many women keep their abuse or harassment secret, further empowering their persecutors. Maria Ontiveros adds that the final crushing knock-out occurs by the legal system. "Once an incident of workplace harassment becomes a lawsuit, the legal system provides the final construct of the event. The legal system's perception of women of color affects cases of workplace harassment brought by these women."[43]

Harassment that develops into sexual assault will end up in the legal system. Based on the traumatic history of Black people with the legal system, African Americans know that the scales of justice are not blind

---

[43] Maria L. Ontiveros, "Three Perspectives on Workplace Harassment of Women of Color," *Golden Gate University Law Review*, vol. 23, no. 3/4, 1993, digitalcommons.law.ggu.edu/cgi/viewcontent.cgi?article=1600&context=ggulrev.

to us and usually tip away from our favor. As a survivor, I know the fear and trauma that occurs beginning with the reporting process, all to end with attorneys, legal briefs, and scrutiny of your personal life, leaving the victim of the assault in shreds, like a horror movie that replays constantly in our minds.

The moment the sexual harassment escalates, when it goes beyond the already-inappropriate light touching, unhumorous jokes, comments about what we are wearing, and when the unwanted advances persist, our only viable recourse is to leave. Black women are three times more likely than White women to experience harassment, which means we are three times more likely to leave because of it.

A May 2020 *Time'sUp* article, "Black Survivors and Sexual Trauma, tells us that the effects of the harassment remain for an extended period of time, impacting our emotional and mental health.[44] If the shift from infrequent, subtle touching or comments shifts to more overt physical touching, or an all-out threat and a sudden departure is made necessary, the trauma does not go away when we leave the company. However, the premature exit from a job could mean a step back in our career or a gap in our resume while we work to find new employment. The collateral damage can be felt financially as well, hurting the livelihood of our families since we are often the primary breadwinners.

> "An estimated three out of four sexual harassment cases are never formally reported. When they are, 75 percent of

---

[44] "Black Survivors and Sexual Trauma," *TimesUp*, 2020, timesupfoundation.org/black-survivors-and-sexual-trauma/.

victims report experiencing some form of retaliation. In addition to the barriers that prevent survivors of other sexual traumas from reporting, survivors of *work*-related sexual misconduct also legitimately fear for their careers, promotions, and even their safety.

Sexual harassment (and sexual assault in the workplace) often drives survivors to withdraw from their work, move, or change jobs, potentially at great economic or professional cost. And when harassment drives survivors out of highly-specialized fields that they are deeply invested in, the career loss can result in 'profound grief.'"[45]

In an article on workplace harassment, Ann Marie Houlis quotes a Comparably poll in which she found that a notable 33% of women executives and engineers, 28% of women in tech, and 34% of women in IT say they've been sexually harassed at work. In fact, 61% of women in IT say that being a woman has even held them back in their careers.[46] I can't help but believe that these percentages are much higher because the poll relies on the women to self-report, and for many prohibitive reasons, a woman still may not want to report abuse even anonymously on a questionnaire. After all, we are scientists, so we know the lengths that

---

[45] Ibid.

[46] Ann Marie Houlis, "Study Finds Black Women Most Likely to be on The Receiving End of Workplace Harassment," *IM Diversity*, Feb. 27, 2018, imdiversity.com/diversity-news/study-finds-black-women-most-likely-to-be-on-the-receiving-end-of-workplace-harassment/.

it takes to make respondent data truly blind so even an electronic survey that doesn't ask for a name can capture a computer's IP address.

These polls also require the respondents to identify the harassing acts that they experienced as abuse and label them as sexually harassing. Comments like "that skirt really shows off your legs today," constant staring at your breasts instead of your face when talking, or inappropriate sexual jokes that are not too graphic but do make your stomach turn all have to be viewed in their full context and mentally connected and labelled as wrong before a woman even considers putting them in the appropriate category of sexual harassment.

Did you know that the 1964 Civil Rights Act was originally meant to protect Black women as well? In 1944, an Alabama woman named Recy Taylor was raped by six White men. Rosa Parks was sent by the National Association for the Advancement of Colored People (NAACP) to investigate the case. Parks' work inspired the formation of the Committee for Equal Justice, which later became known as the Montgomery Improvement Association. The social movement, widely described as the Civil Rights Movement, emerged out of Black women demanding control over their bodies and lives and Black men being killed for protecting Black women, which is ultimately the fight for Black women's bodies against White supremacist rape and assault.

Decades later, Black women still need protection from sexual violence, despite the work of the Civil Rights Movement. The U.S. Equal Employee Opportunity Commission states that it is unlawful to harass a person (an applicant or employee) because of that person's sex. Harassment can include "sexual harassment" or unwelcome sexual advances, requests for sexual favors, and other verbal or physical harassment of a sexual nature. Harassment does not have to be of a sexual nature, however, and can include offensive remarks about a person's sex.[47]

For example, it is illegal to harass a woman by making offensive comments about her in general. Although the law doesn't prohibit simple teasing, offhand comments, or isolated incidents that are not very serious, harassment is illegal when it is so frequent or severe that it creates a hostile or offensive work environment or when it results in an adverse employment decision (such as the victim being fired or demoted). The harasser can be anyone, even an external vendor.

Sexual harassment training is now mandated across every tech company in America, and EEOC policies are clear—persistent gender offensive comments and jokes can create a hostile work environment. But are instances of sexual harassment decreasing? Title VII of the Civil Rights Act of 1964 includes language that discrimination based on sex is sexual harassment in the workplace. However, did that send a message to victims that they had a right to speak up, be heard, and that something was going to be done to eliminate the behavior that was occurring? Did it send a message to those in power that if they make threats, sexually oriented jokes, or inappropriately touch women colleagues that they should stop or risk being fired or sued? Not so much. The research shows that while women may have at first found an avenue for help, retaliatory practices and a culture of keeping quiet took

---

[47] *U.S. Equal Employment Opportunity Commission*, eeoc.gov/sexual-harassment.

hold, resulting in the decline in the number of women who report workplace sexual harassment.

The U.S. EEOC has seen a decline in cases overall from 2018, but Andrew Keshner attributes the Me Too movement, initially started in 2006 by Tarana Burke and amplified by actress Alyssa Milano in 2017, to the EEOC receiving 7,609 complaints specifically regarding sexual harassment in 2018, which marked an annual increase of nearly 14%.[48] Perhaps the movement has infused a renewed faith in employers to do the right thing and make actual, lasting change to shift the workplace into being safer and more inclusive.

Three months as "the new girl" on my team, my teammate, a Black man, took to giving me unsolicited neck and shoulder rubs while he stood behind me. He often told me to "loosen up" and stop being "so serious" and that I was too stressed and needed to relax more. He was a contract developer, not even a full-time employee of the company, so his demeanor and "cool guy" attitude were well received by the team. As the newest and youngest team member, I certainly didn't want to rock the boat of the team dynamics. When I inquired of my cube mate, a slightly older woman from Sierra Leone, if he massaged her, she said, "Yeah, sometimes," with a squirm and shoulder shrug and put her head down. That was the informal message to me to bear it and ignore it.

The pattern occurred every Monday, Wednesday, and Friday during the prep for a status report of our

---

[48] Andrew Keshner, "Why workplace sexual-harassment complaints keep climbing," *MarketWatch*, April 25, 2019, marketwatch.com/story/why-workplace-sexual-harassment-complaints-keep-climbing-2019-04-25.

joint project to our supervisor. He would stand behind me and proceed with a massage that literally made my stomach do flip-flops. If I stood up, he would instruct me to sit down. If I told him I felt fine and the massage wasn't necessary, he would shrug it off and proceed anyway. After all the subtle and non-verbal cues I'd used to indicate that I was 100% uncomfortable, I decided to cringe my way through the prep session as I watched the minutes tick by on the lower righthand corner of my computer screen until the meeting concluded. So, I sat there, tolerating his touch, bearing his harassment, in a room full of people, every Monday, Wednesday, and Friday, until his contract was over six months later.

It was 2005, about two months before my wedding, and I was preparing for a one-on-one meeting with my supervisor—the same supervisor who refused to hear a project status report directly from me, even though I was project manager and that was my job. This time, the topic of discussion was supposed to be my DAP, or Development Action Plan. The DAP creation occurred in January and was reviewed by July. While I was fully involved in my wedding planning, I came to the meeting not with wedding plan updates but fully prepared to discuss my skills, interests, and next professional move onto (prayerfully) another team. I was armed with an agenda, notebook, and pen, but instead of the career-planning and training update I expected to come from this meeting, what I got was a lecture in abstinence! Yup, you read that correctly. . . abstinence!

My supervisor explained that he was a couple's counselor for his small church in Nowhereville, Indiana with his wife. He said he was very interested in ensuring that couples start off on the right foot for marriage. He lured me into the conversation by asking about my wedding plans and where my fiancé was living until the ceremony, and his questions became more and more intrusive when he finally dropped the bomb. "Angel, it is extremely important to remain faithful and pure until marriage."

He kept talking, but I honestly didn't hear anything else he said. I faded away into a cloud of utter disbelief. Aside from the fact that this man filled my day with microaggressions and micromanagement tactics while completely ignoring me when I spoke during the team project update meetings and interrogating me whenever I returned from lunch, asking me where I went to lunch and with whom, aside from me absolutely distrusting him with every fiber of my body, I had no, I mean ZERO inclination, to discuss my sexual status with him. . .ever! His inappropriate interrogation made me mad. My madness built to rage. In my head, I said, "Enough!"

I sat straight up, and with my stern "angry Black woman" look and tone, I said, "I have a counsellor and my own church's couples pastor, whom we see regularly, and those topics are personal and will be discussed with him. Now, I think we should get back to discussing my career development."

He smiled and replied, "Oh, good. Sure, sure, I'm glad you have someone. Well, if you ever need to talk, that's what my wife and I do."

That's when my righteous indignation turned to fear. My "angry Black girl" look didn't faze him one bit. He didn't recognize that he'd crossed the line. *Sweet Jesus, what now?* I wondered.

For these situations, you call on your tribe—a network of professional and personal friends and mentors who can listen, offering sage advice, wisdom, a moment of empathy, and practical steps of how to navigate the situation. I needed to know how to engage with this nut-job in a way that wasn't going to jeopardize my career. I, however, did not have those people at that moment. My emotions took over and had the run of the house. By the end of the day, I was literally seeing red, and in the weeks that followed, the red turned crimson. I avoided this man at all costs. I ignored his emails, picked up the phone when he walked past my cubicle, and set up meetings in other buildings.

Our next set of private one-on-one sessions were conveniently double-booked with important customers. Then I got the bright idea to talk to HR. My plan was to plead to be moved to another team, take another role if I had to—anything to get out from the nightmare. Well, a trip to HR wasn't an option, as my company's policy was that all issues between an employee and supervisor first be handled with a consultation meeting.

The meeting went like this: someone from your same gender, ethnicity, religious, educational, and socio-economic background would pair with you, and your counterpart would receive their pairing. The individuals with the issue and their pairs would sit down for a good old-fashioned mediation. The pairs were hand-picked for their high emotional intelligence and trained on techniques that helped them spot and address unconscious bias and resolve miscommunication between others that typically result from cultural, racial, ethic, and gender differences. The goal is to address the elephant in the room in a safe space and on neutral ground.

Sounds pretty good, eh?

Ha!

After all that training, 90 minutes later, both pairs realized that this guy just wasn't getting it. The best thing that came from the session was my vindication that I wasn't crazy, and at least two other people heard firsthand what a nut-job my supervisor was. Both stated openly that his behavior was beyond inappropriate, unprofessional, and should be formally reported. The session ended with a threat to report him if he didn't cease and desist immediately.

I will say that the program did a great job of selecting pairs because they found someone just like me; by the end of the meeting, she was so angry, she stood up and leaned on the table and used her best "angry Black mom" voice to tell him that offering an employee pre-marital counseling during a career development session was without a doubt crossing the line. When he flashed the same "little boy caught playing in dirt in his dress clothes after church" look toward her, she yelled, "Enough!" and threw her arms out on either side like an umpire yelling "you're safe" to a baseball player.

Her counterpart, who was paired with my supervisor, was red as a beet and moved his chair away from the guy. I interpreted his physical distancing of himself to mean, "I'm not like THIS guy."

My pair gathered her paperwork, pushed in her chair, took a deep breath, and ended with, "You just can't do that. It's not your place, and you must keep conversations with her professional. If you don't, I'll report you to HR myself...do you get it? Are we clear?"

He hung his head like a scolded toddler and slowly shook his head up and down as he folded his hands and put them on the table. He sagged down in the chair and mumbled, "I was just trying to help."

Both pairs looked at each other in utter disbelief, and my sister girl pair looked at me with an unspoken "good luck, girl." Her last sympathetic look toward me marked the moment my hopes of justice went up in smoke. It was evident to the three of us in the room that he wasn't capable of understanding that it was beyond unthinkable to ask an employee if she was having sex with her fiancé. So, since he didn't get the message on that topic, then there was zero chance in hell of waking him up to the fact that his daily microaggressions and micromanagement were pushing me into a depression. The constant and consistent barrage of questioning my every move was humiliating and oppressive, and after our failed mediation, I knew it wasn't going to stop.

This is one of those stories where it gets worse before it gets better. He discontinued the sex talk, but the micromanagement remained just as before. I would double- and triple-check every email, started doubting my own decisions as a project leader, and would hold at least two meetings with my peers (fellow project managers) to ensure my logic was sound before I presented the project plan and schedule to management for approval. All this extra work was not only emotionally draining but also time-consuming. I couldn't carry as many projects as my peers due to the added burden and stress. So, not only was "nut-job" screwing with my mind and emotions, he was negatively impacting my work performance too.

I suffered in silence for a full year and half after the "pairs" showdown. I slowly reached my boiling point and finally snapped after he told another manager that I was a slow worker and difficult to manage. I made an emergency appointment with my HR rep for later that day. Sitting outside her office, I waited for her to call me in. When she opened the door I raced in and let it ALL out. I told her about the "pairs" meeting, the insults about me getting married, the pregnancy jokes, and ignoring me in meetings. I managed to emotionally vomit two years of pent-up suppression in less than ten minutes. I ended with, "Amy, it's either him or me!" I shouted, "My resume is updated, and I'm hitting the send button to every job recruiter in town when I get back to my desk!"

Now keep in mind, I didn't expect her to stop me. She was our HR rep who I really didn't know very well, and I was a mid-level project manager that was easy to replace. I certainly didn't consider myself special or unique to the company. I just wanted to unload the pent-up pressure for some immediate emotional relief. I wanted to be unyoked from the crushing weight I'd been carrying around and make it clear as to why I was leaving. Honestly, I just wanted to expedite my exit interview, but what she said next knocked me back into the chair by the door and left me slack-jawed for a full minute.

She said, "Angel, calm down. You aren't crazy. It's not just you. Please don't quit. We know all about him. We are doing an investigation on his conduct."

She went on to ask me to give her a little more time, and she promised the situation would be resolved. Exactly 30 days later—I know because I counted—our entire team was called into a conference room, where it was announced that he was leaving. The team was to split in half, with half going to another team and my half going to a new supervisor. A bit of fear started to rise in my chest until the new supervisor walked in—we will call her Jamie. Jamie was one of the toughest, smartest, most fair and supportive and encouraging managers that I have ever met. The good news kept coming—a new Senior Director was coming to our department as well, the same advisor and coach who sponsored me during my internship! Let me tell you that at the conclusion of that meeting, my teammate, Lynda, and I danced all the way down the hallway back to our desks. Margaritas and tacos were on the dinner menu that night while the dancing and celebration continued at home.

Even though the new leadership projects were still tough, business partners were challenging, coworkers weren't always agreeable or willing to help, and the long nights and consistent need to prove myself didn't help my new marriage—life wasn't perfect, but the support and "you can do it" talks from Jaime taught me lessons I remember to this day. While the job didn't change (in fact, it got harder), having a boss who didn't devalue my work in front of other people, make pregnancy jokes, or try to get me to talk about how much sex I was having was as different as night and day!

Frontline Tales

> As an IT Manager, every time we had a lunch meeting, my supervisor would open my door, make small talk about my family and daughter, and constantly ask if I was okay; it was nice on the surface, but it always made me feel like we were on a date.
>
> —Hannah, IT Director

> I remember a time when I was on the elevator, with about ten Caucasian men. So, the first stop on the elevator was me getting off. And when I got off, there were a few hands that were on my bottom.
>
> There were some senior people in there, so I didn't look back. I just kept going, but I felt extremely violated.
>
> —Laura, Senior Program Manager at Multinational Engineering Company

"Oh, do you want to get ahead in your career? We should talk." And he became a little more aggressive to where it became a situation where it was about my willingness to be flirty. And then I was told, "If you're willing to do a little more, I can make sure that your career flourishes." I told him that I wasn't willing to do that, then there were consequences for that. So, I lost my job.

—Stephanie, Entry-Level Associate

[Myra was meeting with the existing Chief Human Resources Officer of her company. Being there for a few years, she was ready to move up and was attempting to implement advice from a mentor as she went looking for a sponsor.] . . . they [her company] had a change in leadership, and our [new] leadership believed in looking at women but not necessarily working with women. [This included the head of HR; and the Chief HR Officer propositioned her by telling her what other women were willing to do.] . . . "and you won't believe what they're willing to do to learn business.". . .And I sat there like, "I gotta go. What do you mean? I'm not willing to do whatever they're willing to do. So, let's end this right here."

—Myra, IT Project Manager at
Global Staffing Company

At two different companies, where one was my supervisor, and then—well, another was a director-level business partner . . . it wasn't real blatant, but it was obvious what they were asking for. It was kind of a, "We need to get together. My wife's not going to be in town." [The other] one would always kind of—because I'm kind of top-heavy, he would kind of look at me but not look at me in the face.

—MaryAnn, IT Manager of Data Sciences

There was this one guy, and he was married to one of the engineers on the floor. He was rude. He was lecherous. But anyway, I tried to stay away from him as much as I could. He was one of the truck drivers on the manufacturing floor. He would say things like, ". . .I knew that was you. I was looking at you from behind."

I had that experience with a supplier one time who came to visit me. I had a skirt with a slit on, and he was talking about my legs. I was like, "Oh my God." And he was one of our own [Black] too. This was when I was younger in my career, these things. I think, to me, some African American men felt more comfortable saying those things to African American women, honestly, because now that I'm thinking back, I can think of a few situations where things were said to me while we're in meetings or even outside the meetings.

—Tina, Value Management and Six Sigma Manager of IT Services

Some situations that occur in the workplace to us are so unbelievable that they are hard to process. This next interview was exactly that, and for this reason, I have included excerpts of the full transcript.

Respondent: So, I've never had this conversation, Angel—and you can decide if you want to include this. I've had a White man kiss me —

Interviewer: What?

Respondent: —on the job.

Interviewer: On the cheek?

Respondent: No, ma'am. Directly on my mouth.

Interviewer: Oh my—why? What context was that?

Respondent: We were standing outside talking—

Interviewer: Okay, and, well, hold on. What was his relationship to you? A customer? Supervisor? Peer?

Respondent: He was my boss.

Interviewer: Your direct supervisor?

Respondent: Yes.

Interviewer: How long had you been working with this person?

Respondent: It was a new—

Interviewer: New relationship?

Respondent: Yeah, it was a new relationship.

Interviewer: And you guys are out—are you in the—

Emotional Misconduct

Respondent: We left out of the office, and we're just in a conversation talking about work. And he just lunges over and kisses me on my mouth.

Interviewer: You come outside. You guys are hanging around chatting it up.

Respondent: And for whatever reason, he reaches over and just kisses me on my mouth. And I was so stunned, I was like, "Excuse me?"

Interviewer: In the middle of the conversation?

Respondent: Yes. I have never—I replayed that over and over again and—

Interviewer: What did he say after he did it?

Respondent: I said—well, I don't know if he thought I was going to do that because he was my boss and maybe he had had some experience with—

Interviewer: Right, a previous situation or—

Respondent: I don't know.

Interviewer: So, you remove yourself from the situation. You get out of there. And then the next day, you just asked to be reassigned?

Respondent: Yeah. I just—I was like, "I'm not going back there." Like, I'm not—

Interviewer: Okay. You still work with the company. You just are now in a different department or...

Respondent: No. I completely left the organization.

Interviewer: You literally just never showed back up the next day.

Respondent: [Correct.]

—Reagan, Senior Consultant of Big Five Consultant Agency

Emotional Misconduct Clues

BYOF

BYOF stands for "Bring Your Own Friend" and requires you to recruit an active ally to help. This person will need to be a trusted confidante who can be available when needed during meetings, on conference trips, or whenever unwanted sexual advances are known to occur. This is not an ideal solution and can not prevent the unexpected pop-up occurrences or surprise encounters, but it goes a long way to provide some psychological support.

Being "the only" and then carrying the weight of being targeted with sexual harassment is too much for one person to carry. Enlisting a trusted teammate to be on the watch or serve as a buddy so you aren't alone does offer some measure of help. A male ally in my department, "Mark," made himself available by adding body-guarding skills to his resume for one of my most professional and competent project manager contractors.

She asked me for an emergency one-on-one session. I assumed it was about issues with her project timeline, but when I sat down for the discussion, she wouldn't. In fact, she started pacing the floor. After a few minutes of stammering and hand-wringing, I learned that another contractor, a guy from a different company, was openly flirting with her in meetings to the point of embarrassment. A quick investigation revealed that one of our employees witnessed the flirtation and agreed that it was inappropriate.

Armed with verification, I was ready to take it to the top of everyone's HR and file a formal complaint to have him removed from the project and hopefully take it a step further and get him fired, but she didn't want any of that. Instead, she insisted that all she needed was for the other coworker, Mark, to be in the room with her and the other contractor. She asked that she never have to be alone with the flirtatious vendor, and if Mark couldn't attend, they would cancel the meeting. This is how they operated. Mark would physically sit between her and the flirt during in-person meetings, acting as a barrier, and any meeting in which Mark couldn't attend would be re-scheduled.

Now, I have serious misgivings about how absurd that resolution was, but it worked for her and she was instantly calmed, which allowed her to perform and complete the project. She never had to work with "Mr. Flirt" again. This is not how I would have liked for the situation to be handled, but simply having others aware that the behavior was occurring and having an active ally physically present as a type of bodyguard was what she insisted upon.

I must reiterate, I didn't agree and reported it anyway to my company. I wanted the flirt held accountable by having the incident put on his formal record in hopes of preventing another woman from being subjected to his behavior, but not everyone is up for a Battle Royale. Sometimes, they only need an ally to be present as a formal witness to feel a bit safer. Despite

Dents in the Ceiling

the inconvenience, by making it someone else's business to be on the watch for when the perpetrator is near, the harassment was eliminated and did not escalate. If you are enlisting someone to intervene on your behalf, I encourage you to document what is going on if there ever comes a time to call in formal reinforcements.

BYE BYE

Leaving is an option. Navigating being "the only," microaggressions, bullying, and being treated like the help is one thing, but once your supervisor or a person in a powerful position thinks that they can touch you, your decision to exit is justified. Staying for financial obligations, for fear of not finding employment elsewhere, or being comfortable where you are should be considered, but the option of leaving must remain on the table.

Financial obligations are legitimate concerns. 29 out of the 30 respondents I interviewed self-identified as the primary breadwinner of their household. These findings support research from a 2017 Center for American Progress study which notes, "A substantial 84.4 percent of Black mothers were primary, sole, or co-breadwinners in 2017, compared to 60.3 percent of Latina mothers and 62.4 percent of White mothers."[49]

We are often not only responsible for the mortgage and kids' tuition, but we often support aging

---

[49] Colin Seeberger, "Nearly Two-Thirds of Mothers Continue to be Family Breadwinners, Black Mothers are Far More Likely to be Breadwinners," Center for American Progress, May 10, 2019, americanprogress.org/press/release/2019/05/10/469660/release-nearly-two-thirds-mothers-continue-family-breadwinners-black-mothers-far-likely-breadwinners/.

parents or family with health care needs, not to mention credit card debt. It reminds me of *American Idol* winner Fantasia. After her win in season three of the show, she admitted she was financially supporting everyone in her family, including able-bodied adults who could have been supporting themselves. The pressure she was under was chronicled in her reality TV show.

Being everything to everyone, including paying their bills, puts Black women in a precarious situation when things get intolerable at work and we need to leave. Just like leaving an abusive relationship, you need to create an exit plan.

Don't wait for the escalation to occur or for the comments, jokes, and suggestions of getting a hotel when the wife is out of town to get to the point of your job being threatened or your physical safety being jeopardized. Now is the time to create a strategic plan to leave the department or leave the company.

Updating your resume, plugging into your personal and professional networks for leads or openings in other areas or other companies, and other practical steps to create a Plan B for yourself are important. Think outside the box, such as teaching and consulting to supplement your income until you land somewhere if you choose to continue in corporate America. This may very well be the catalyst for you to strike out on your own and build your own practice, which is certainly an option many of us choose. You can further get a handle on your financial situation by down-sizing your expenses to keep afloat until you can make your next major move.

You are enough. You deserve it. You can do it. You do have technical aptitude. You can learn new skills. You can rebrand yourself in a new department or at a new company. You can move. Tell yourself this repeatedly until you believe it. Do not be fearful of moving, no matter how long you've been at your current organization. As the scripture says, "God did not give us a spirit of fear." It is a lie that is keeping you stuck.

The golden handcuffs are real—I know, I wore them for years! My biggest fear was that I didn't want to have to learn a new devil. By this, I mean that being at an organization for over a decade, I knew the snakes, I knew which senior leaders to stay away from, which ones were allies versus enemies versus indifferent; I knew the culture, the history, the good and the bad. I was so scared to have to learn all of that all over again and to prove myself again. The original battle to obtain all that cultural knowledge had been such an uphill climb I was not looking forward to repeating it. However, when my supervisor asked if I was having sex with my fiancé, all that fear flew right out the window, and I was ready to re-learn whatever I needed to just to be free.

Your freedom from harassment is paramount. You do not have to suffer in that environment. You have options, and if fear is paralyzing you, I encourage you to seek a mentor or perhaps a professional counselor to help you work through your fears to take charge of your life and move from surviving to thriving. Remember, you are enough. You deserve it. You will do it.

Legal Eagle

Remember all the steps you took to avoid bullying in Chapter Two? This is that on steroids! Get a calendar—a calendar on your phone or home computer is best (not your company-provided cell phone) and capture the time on the calendar when the incident occurred. It could have been a sexual joke, an unsolicited neck rub, an insinuation that if you don't comply with an act that your performance or job be negatively impacted. Whatever occurred, write it down! Even if it is minor to you at the time, it paints a completely different picture when those minor events are viewed together over several months. The key to proving harassment has occurred is to demonstrate that the acts were persistent and pervasive enough that a reasonable person finds them offensive.

Consider consulting an attorney. Your situation may warrant some legal expertise especially since these types of situations are usually difficult to prove. Putting all these clues together as documentation, focusing on increasing your confidence and creating a solid exit strategy, enlisting allies, and consulting with a legal expert will help you get away from an environment where someone can't seem to keep their hands to themselves.

Active Allies Against Sexual Harassment

Secret Agent Man

I know that you didn't think a career in tech could lead to going undercover, but being an active ally may call for a few unconventional tactics. Offering your ear to listen to someone is great. Stepping up to sacrifice your time is the next step.

Once you see sexual harassment, pretending like you didn't, or worse yet, turning a blind eye is not the way. One way to stop it may simply require your presence. If you know the harasser is getting close, inserting yourself into the physical space could help.

For example, as the other team members were standing around our weekly team meeting, and I noticed how odd it was that I was the only one sitting down with the contractor's hands on my neck, I figured it must have stuck them as odd, too, to say the least. One of the other team members just planting themselves between me and the contractor could have been enough to stop the unwanted massages.

Instances of sexual harassment are not always so blatant or overt. For the situation in which you are told about but haven't witnessed, you are now on notice. Note the days/times in which the accused harasser is around the woman who told you; document when you see them together even if you didn't hear or see anything inappropriate. Your information can go toward her credibility since these situations often come down to a he said/she said dispute.

Go Tell It on the Mountain

This is the equivalent of "see something, say something" advertisements for workplace violence, which is appropriate since, to women, sexual harassment very much feels like workplace violence. If you find yourself popping up like the butler in Adam Sandler's movie *Mr. Deeds* whenever your female coworker is being approached by the person she says is harassing her, then it's time to mount your testimony.

Active allies should consult with professionals to prepare the documentation they need for formal reporting. The steps that I shared above regarding the documentation that she will have to follow are steps you will have to take as well. Observing the harassment will go a long way, but writing down what you saw or overheard and when it occurred will be beneficial when the time arrives for you to file a formal report. Kayla Matthews provides considerations in preparation for filing a formal grievance:

"If you witness sexual harassment in the workplace, who do you report it to? The decision ultimately depends on the size of your organization, your level of trust, and alternate reporting options.

In general, beginning by reporting in-house proves most effective, especially in cases of a hostile work environment. This allows HR representatives to work privately with offending individuals to modify their behavior. Proceeding in this manner can ease interoffice relationships, as the party need not lose their job if they display genuine remorse and a commitment to change.

However, if you work for a small firm, things may be more difficult. There may not be anyone dedicated to processing such claims. In this case, file a complaint with the Equal Employment Opportunity Commission. ... Finally, you'll want to contact a licensed employment attorney."[50]

If I Was Your Girlfriend

If you are a male ally, think about your girlfriend, wife, or daughter spending time alone with their supervisor and being touched and harassed against her will. You certainly wouldn't want her in that situation. Instead, you most likely relish the thought of meeting

---

[50] Kayla Matthews, "Steps to Identify and Prove Sexual Harassment in the Workplace," *Power to Fly*, Sept. 4, 2019, powertofly.com/steps-to-identify-and-prove-sexual-harassment-in-the-workplace-2640185061.html.

the culprit in a dark parking lot. If you knew that a coworker witnessed her harassment, you would absolutely want that person to support her and report it. Now, envision you are that person. . .what will you do?

Jeffrey Tobias Halter is a leading expert in DE&I and a gender strategist for engaging male advocates in the workplace to advance women. He initiated the *Father of a Daughter Initiative* where male allies can pledge their support of the women that they work with to support them, advocate for them, and actively promote them. Halter postulates that men tend to compartmentalize their lives.

He self-reported that while at work, he was focused on completing his daily activities, going to meetings, and sending emails, but never once looked critically at his organization to see the disparities in hiring, promotions, or how women and minorities were being treated. He started paying more attention to how women were treated at his job once he had his daughter. He missed all the times that women were talked over, ignored, regulated to supportive work, touched, or harassed. Thinking about how the work world is versus how it should be became more apparent to him once his baby girl arrived.

He and many of his male counterparts aren't being malicious; they truly aren't seeing what is going on, and if they do, they aren't sure what to do about it. Halter woke up to the reality that his silence and compliance was fostering a workplace environment that was not inclusive. It was similar to how the resurgence of the Black Lives Matter movement in the wake of the death of George Floyd made some White people aware of their own blindness toward racial injustice. If Halter had any hope for ensuring that his daughter and everyone's daughter could safely start their job free from stereotypes, persecution, and discrimination, he and now you will have to work hard to make that a reality.

I spoke with several women who reported being supported by allies. The role of advisor, mentor, sponsor, and how they help advance women in the workplace is very important, but the role of an ally is unique. Active allies can show up anytime and anywhere without being asked. As found in this next example, the woman was shocked when she was pulled into an HR meeting in which someone else had reported her being treated unfairly. This was the brave act of an extraordinary ally.

> I think that one thing I'm noticing is we're in an era of change. And so, I'm in this HR conversation. This woman [the HR rep] is like, "Someone has reported that someone was yelling at you or that they are saying these things to you. You're one of the only women Agile coaches here at your level. Do you feel like you get discriminated against?
>
> —Vivian, CEO of Agile Coaching Practice

Your mission as an active ally is to show up just like the person who reported their colleague yelling at this agile coach. If you see something, say something, especially when verbal and physical abuse or harassment is occurring.

Breakthrough Tools

- Sexual harassment truly begins as emotional manipulation and quickly moves to abuse. It resembles what was described in the chapter on bullying. Remember: it is about control.

- If the person is making you uncomfortable, listen to your instincts when your body is screaming warning signs.

- Human Resources is not your only option. Do not be afraid to seek legal counsel.

- Do not stay stuck. You can start over. You can build political capital again. You can be successful and thrive at another company. It's your body and your career—do not hand it over to anyone. The longer you stay in a toxic work environment, the more difficult it will be to show up healthy when you have to start over elsewhere.

Starting Points

Women: Document your experiences. Keep a calendar or notebook and track the harassing incidents.

Allies: Again, pay attention! Be mindful and alert of what is going on with your team. Inquire to see if your team members are uncomfortable in any work situation and could use help. Being open and getting to know your staff puts you in a position to learn what is really happening on your team so you can intervene.

Chapter 6

Macro Hurts

"I changed what I could, and what I couldn't, I endured."

—Dorothy Vaughan,
Mathematician

I think the saying is "death by a thousand paper cuts." Daily attacks on your character, competence, and self-esteem can seem harmless on the surface, however, they always take a heavy toll over time. The microaggressions mount so high that they take their toll emotionally, causing job performance to suffer, which again leaves very little recourse with many women choosing to suffer through it or leave. This is the experience of when microaggressions turn into macro hurts.

The term "microaggression" emerged in the 1970s and is best defined as "a comment, action, or incident regarded as an instance of indirect, subtle, or unintentional discrimination against members of a marginalized group such as a racial or ethnic minority."[51] Most microaggressions take place in the form of subtle comments such as, "Oh, well, I won't get into the details and take up your time explaining this complex piece of code. I know girls aren't that technical."

Depending on the tone and context, when this comment is delivered with a smile, it can seem almost considerate. You can assume the speaker is trying to be nice by letting the woman he is addressing "off the hook" from being embarrassed about not knowing or understanding complex, technical terminology or concepts found in many computer systems. Unfortunately, this comment is anything but harmless. It is filled with bias and unsupported assumptions. Instead, inquiring what her foundational knowledge or background is in coding would be a better start, and asking if she is interested in learning more is preferred.

Furthermore, the initial assumption robs the woman from a great opportunity to learn if she is not familiar with coding concepts. Lastly, what experience does any man have that leads to the wrong and sexist assumption that women aren't technical? The first email was sent by a woman. NASA employed a whole team of human mathematicians who were women to check the computer's calculations before they were accepted. And it was Dr. Valerie Thomas who invented the technology that made 3D imaging possible.

Jonathan Kanter, a clinical psychologist and director of the Center for the Science of Social Connection at the University of Washington, uses an iceberg analogy to explain how microaggressions fit into the bigger picture of prejudice. The tip of the

---

[51] "Microaggressions," 2021, oxfordlearnersdictionaries.com/us/definition/english/microaggression.

iceberg is overt racism, sexism, or homophobia, which is visible and unmistakable. Microaggressions are the harder-to-see biases that lurk under the surface, more common than overt racism but less detectable. "The sea the iceberg floats in is the bias enabled by society and institutions."[52] There is so much to unpack in just that one comment, "Girls aren't that technical," but the real issue is that dismissive inferences like this are made daily. It would be exhausting to confront each one. Like a marriage, a work relationship requires us to pick and choose our battles.

Comments and intrusive gender, racial, and/or ethnic questions are annoying at best and offensive at worst. Microaggressions take on a more deceitful shape when our work performance is impacted. It is one thing to assume I'm not technical because I'm a woman, but we have just kicked it up a level when it's in my performance review that I'm "not technical enough."

Huh? Based on what grounds? Can we define what "enough" is for this role? Is there objective, written criteria provided so I can find examples of how I meet the criteria or create a development plan for how I can work toward increasing my technical acumen?

Without answers to these questions, we can quickly see the challenges that African American women experience when we are constantly being ignored in a meeting, or our ideas not taken seriously when we are not given the financial support or time to be developed compared to our counterparts. Even having the bar set too low for us because it is assumed that we are not competent or do not have the capability of handling challenging assignments diminishes our

---

[52] McKenna Princing, "What Microaggressions Are and How to Prevent Them," *Right as Rain by UW Medicine*, Sept. 3, 2019, rightasrain.uwmedicine.org/life/relationships/microaggressions.

chances for promotion. African American women in corporate America have been the subject of many studies around the topic of microaggression since it is so prevalent among our population in the workplace. After the term was coined in the 1970s, it has been increasingly studied since the 1980s. We also know from qualitative research conducted by Solórzano, Ceja, and Yosso in 2000 that African American students who attend Predominantly White Institutions (PWIs) experience the undertones of being inferior to their White peers, considered not good enough, excluded from valuable study groups, graded differently, and treated with bias by professors.[53] These negative messages can easily date back as far as grammar school depending on the racial, ethnic, and gender composition of the school and neighborhood.

We must think about the deleterious emotional and psychological effects microaggressions have on us as human beings. The erosion of our confidence, courage, and resolve amounts to absolutely nothing micro at all but a "macro hurt." It first takes awareness that affliction is even occurring because often we don't see it for what it is due to its insidious nature, but when the revelation is made, it then demands intentionality and hard work to counterbalance its effects. Such psychologically ill consequences result in what we have now come to know as "emotional tax."

Catalyst, a nonprofit organization known for its work in accelerating women's careers in leadership roles, defines emotional tax as "the combination of feeling

---

[53] Daniel Solórzano, et al., "Critical Race Theory, Racial Microaggressions, and Campus Racial Climate: The Experiences of African American College Students," *The Journal of Negro Education*, vol. 69, no. 1/2, 2000, pp. 60-73, http://www.jstor.org/stable/2696265.

different from peers at work because of gender, race, and/or ethnicity and the associated effects on health, well-being, and ability to thrive at work."[54] This emotional tax shows up in many ways, but the most common are the need to speak differently, wear modest clothing or darker colors than your personal preference, take extra time to re-read emails before sending, over-preparing for presentations, and mentally churning over what you should or should not have said in a meeting. All the extra time and doubting may seem pointless, but it has all come from learned behavior. We, as African American women in the male-dominated halls of tech spaces, have quickly learned that all these extraneous steps are indeed required for everyday communication.

To this very day, I remember when a brilliant, wonderful woman pulled me aside and gave me my first corporate America hack. She was in her late thirties, had been employed with the company about five years at the time of our meeting, and was a devoted wife and mother of two boys. She was Chinese and spent most of her secondary formal education in Australia, where she met her husband, got married, and then they eventually made their way to America. She was on a work visa and he was on a student visa to pursue his doctoral studies. She learned a great deal in her brief time in the states and decided I was worthy enough to depart her hard-won lessons upon.

I have never forgotten Lesson #1: SAVE EVERYTHING! She literally pulled my arm and stood on her tip toes when she said it and gave me the "angry mom" look. So, when she said it that way, it certainly got my attention. Thanks to her protective advice that I

---

[54] Julia Carpenter, "The 'emotional tax' afflicting women of color at work," *CNN Money*, March 5, 2018, money.cnn.com/2018/03/05/pf/emotional-tax-women-of-color-at-work/index.html.

never forgot, I saved myself from quite a few "I didn't say that" or "I don't know where you got that idea from" situations. Knowing that you must do extra to cover your ass is helpful for our White counterparts but mandatory for us.

Throughout the day, you might find yourself bracing for insults, avoiding social interactions and certain informal gatherings, or adjusting your appearance to protect against hurtful situations. You live each day in a constant state of being "on guard." Catalyst goes on to report that "emotional Tax can deplete Black employees' sense of well-being by making them feel that they have to be 'on guard,' disrupting sleep patterns, reducing their sense of 'psychological safety,' and diminishing their ability to contribute at work."[55] This proves that the cost of armoring up is commanding a higher price than we have resources to pay.

Armoring is one of my personal favorite coping mechanisms. I was taught to armor by my grandmother, who could have won a gold metal had it been an Olympic sport. Armoring is such a part of my DNA that I didn't know it had an official name until I picked up the book by Bell and Nkomo titled *Our Separate Ways: Black and White Women and the Struggle for Professional Identity*.

The book included an in-depth study of how 120 White and Black women clawed their way to reach senior management levels in corporate America. Their study included women from marketing, HR, and sales.

---

[55] Dnika J. Travis, et al., "Emotional Tax: How Black Women and Men Pay More at Work and How Leaders Can Take Action," *Catalyst*, Oct. 11, 2016, catalyst.org/research/emotional-tax-how-black-women-and-men-pay-more-at-work-and-how-leaders-can-take-action/.

They found how the Grand Canyon-like divide was built between Black and White women and what to do to resolve it. A critical theme they explored was armoring.

> "Armoring is a 'political strategy for self-protection,' whereby a girl develops a 'psychology resistance' to defy both racism and sexism. We believe the armoring process is a critical element of the Black woman's psychosocial development."[56]

It was interesting to learn that armoring included many different facets, and that aspects of self-protection manifested itself in a couple of different ways. For example, self-resilience became a noteworthy aspect of armoring in which we as Black women essentially become mentally tough—by the way, I absolutely crack up over the new mental tough coaches who have hit the scene lately. We as Black women don't have to be taught to be mentally tough; it is a foundational element of our being just to get through the day.

Self-resilience is about not relying on another for support. Suffering from too many failed support systems already, the idea that "no one is going to make me fail" plays out at work in which we as Black women sometimes become a one-woman show to do it all and ensure that no ball is dropped on a project or assignment.

As a PM, if I had a deadline for which testing needed to be complete by a certain date, and the tester could not or would not complete it on-time, guess who

---

[56] Ella Bell Smith and Stella M. Nkomo, *Our Separate Ways: Black and White Women and the Struggle for Professional Identity*, Harvard Business School Press, 2001, pp. 96.

was logging into the system to execute test scripts? When a fellow PM found out, he replied that there was no way he would do that and would escalate the situation to that person's supervisor instead. My thought was, why bother when it would get me nowhere and waste time when the testing needs to get done now? The risk of not preventing a project from being completed on time wasn't an option for me. For him, not an issue, but for me, very much a concern.

Another characteristic of armoring is silence. I cannot emphasize the amount of promising that was necessary for me to do to get the women I interviewed to agree to speak with me, and they were still reluctant even after I guaranteed their anonymity. Some ladies required access to their interview recordings, and at least three ladies called and texted me several times to ask repeatedly for their personal and company names not to be shared. For every woman I contacted who I didn't know personally, two turned me down for fear of sharing their story.

One lady was kind enough to explain why she couldn't speak with me. She did not want to recount the microaggressions and acts of blatant racism and sexism that she endured because it would take her "back there," and she couldn't "afford to go there again." I explained that recounting my incidents was therapeutic for me and that it may help her to talk with another African American woman. After a pregnant pause on the phone, she replied, "I'm sorry. I wish I could help you. You are doing great work, but I just can't." Microaggressions have a high correlation with being "the only" or being ignored and overlap with the subcategories such as "White female-missed allyship" and lack of sponsorship and failed allyship, which we will explore later.

One woman who reluctantly agreed to speak with me began the interview by saying upfront that she hadn't been discriminated against and wouldn't make a good participant. Then she exclaimed, "Oh, wait, there is this one incident that happened to me, but I just moved on from it and try not to think about it." As you will later read, it was one of the most overt forms of racist and sexist discrimination that I heard from all 29 interviewees... and she'd forgotten it happened?

I held my first summer internship right out of high school when I was 18 years old. I was able to start interning so early thanks to a program called INROADS. INROADS' mission is to increase the amount of minority representation in corporate America. A dream of founder Frank C. Carr to make a difference, INROADS originated in 1970 with just 25 students and 17 sponsoring companies and grew to an international organization with more than 28 offices serving nearly 2,000 interns at more than 200 companies. INROADS taught me at an early age how to navigate the informal rules of corporate America as a young Black girl.

During this first internship, it happened that the training class I needed to attend was in Indiana and was going to last three full days. I didn't own a car. In fact, my first car ever was a cute little blue Honda Civic, fully loaded, which I got right after my 22nd birthday. The company I was interning with that summer, decided they would rent me a car from a local car rental shop. That was a great plan until I got a ride to the rental place and the manager informed me that they couldn't rent a car to anyone under 25. I called our department administrative assistant (AA), who was incensed!

However, no amount of pleading, threatening, or negotiating was going to bend the rules for me.

I stayed in that rental shop until the AA had called all over the city to secure a car for me. More than an hour later, we found one, and the company had to pay an arm and a leg in extra insurance for me—so no pressure to not get into an accident over the next few days. The Senior Manager directly responsible for hiring me as an intern showed up to transport me from one car rental facility to the other. After she signed in blood a mountain of extra paperwork since I was underage, we returned to the company. We stood at the AA's desk, having a good laugh at how the AA had to perform backflips to make a simple car rental happen, when the Senior Manager turned to me and said, "You really surprised me. You showed more patience than someone like you normally would."

I was a proud intern for three years until a break in a personal relationship forced me to rethink my life. More running from someone but not to anywhere in particular, I decided to spread my wings and try an internship in a different city—any city, really. I was encouraged to interview for a large organization in the Midwest that was not high on my priority list, since I was trying hard to get away from that entire region of the country, but I relented and landed the job. This was the summer of my fifth internship in tech.

The standard process for this company was that all interns gave a mid-summer and end-of-summer presentation of their assigned project. Mid-summer, 15 interns gave presentations, all with "softball" questions throughout. Half the interns even received helpful advice to improve their summer-end presentations. I, however, got grilled for an uninterrupted 10 minutes on technology, processes, and the people I interviewed for my assignment, all of which I had never even heard of before! The interrogation only ended after my supervisor stepped in to cut off the inquisitor by saying, "That was outside the scope of her project." He pulled me aside after the cross-examination and said, "Don't sweat it. That jerk is always tough on women."

By the end of summer, I nailed the presentation, since I was prepared for the Spanish Inquisition. An offer was made for full-time employment, and I accepted despite Mr. 100 Questions who was over-the-top in trying to poke holes in the little Black girl's presentation. I went on to become a learning system administrator and business analyst.

Our company instituted flex time, so for three years, I worked longer hours Monday through Thursday so I could take off on Fridays after lunch. My long-distance relationship took place Friday evening through Sunday, so the standard 4:15pm flight out of Indianapolis every other week became routine. Everyone knew my schedule except my new supervisor who I started reporting to just one week prior. He scheduled a 1:00 p.m. meeting with our vendor, business partner, and myself. Not wanting to be rude, I accepted the invite, even though I knew that I had a hard stop at 2:00 p.m. to race to the airport to barely make my flight. I proceeded to pack up my laptop and gracefully exit the room with my roller bag, wishing everyone a wonderful weekend after my portion of the meeting concluded.

Monday morning, he came to my desk to admonish me in front of our entire team, shouting to never leave one of his meetings early again and that my behavior was unacceptable. He bellowed that I was disrespectful and that I embarrassed him. That very Friday, Tony, a team member, left in the middle of another long meeting for a friend's wedding. I asked my new supervisor if flex time was reinstated, in which he replied, "Oh, that's different. Tony is in the wedding." Lady luck was on my side, and a short three months later, our department was reorganized. Since my position was unique and could fit anywhere, I got the one and only opportunity in my career to select my new supervisor.

After a rough interview for a summer internship, in which I knew I wasn't going to get the offer, the company called me, and much to my surprise, they wanted me to start a week after finals. About three weeks after starting the new gig, the Principal Director who was conducting our weekly one-on-one sessions said, "I bet you're wondering why we chose you." He went on to say, "You weren't the smartest candidate; I mean, you didn't have the best grades, but you're likeable, and since we need to win more business with clients, we knew we needed a female since we're a bunch of guys in this office, and this client is big on diversity." Yes, it was a shot at my self-esteem, but of course it pushed me to prove that though they hired me to check a box, they were going to get much more. I didn't win them that client, but I did secure Excel sheets full of leads for their pipeline that guaranteed them at least three to five more years of business while single-handedly revitalizing their reputation with our biggest customer. I mean, if likability is an asset, why not use it?

It's hard for women, specifically for women of color, when we're talking about getting to the C-suite. Simple logic follows that if no one that looks like us is already in the room to pull us into the room, we need those who aren't like us to open the door. This concept is clearly visualized in a scene in the movie *Hidden Figures*. In the scene in which Taraji P. Henson's real-life character, Katherine Johnson, is frustrated at not having all the information she needs to do her job, she sits outside the big planning meeting filled with White men to hopefully get in the room to hear what is being discussed. All so she can have enough information to create proper flight plans for the space flight. My favorite part of the film is when the head of the department, Kevin Costner's character, Al Harrison, finally opens the door and says, "Let her in!" Yes! This is exactly what we still need—an open door!

Now, many of us are tired of waiting (and don't have to), so we have started to create our own tables and open new doors by establishing our own tech companies. However, there are many of us still along various rungs on the corporate ladder in traditional companies not of our own making, and we're still working our way up. We must navigate the course from an associate or colleague and eventually make our way to a trusted work partner. The mighty step before a partner is "work friend." This is when we make it into the "likability zone" and our "work friend(s)" can get comfortable enough with us before they even consider us a viable candidate for entrance into the C-Suite.

Remember, outstanding performance is par for the course. While penetrating the "likability zone" breaches Similarity Bias, a key differentiating factor when it comes to moving up the corporate ladder toward the C-Suite, entering that zone will require navigating the land of microaggressions.

Frontline Tales

And then I think at work, I've been pretty fortunate to not experience anything that I would say is "crazy." I mean, there's your typical things of people, of certain racists, touching your hair, and you have to have a conversation with them. I have a sign outside my desk that actually says, "Please do not touch my hair."

Sign Reproduced by Jasmine Ballard

I've actually told the CEO that. I was like, "I had to put a sign on my desk. I just want you to know." Yeah, what I had to break it down to was, "I feel like you're petting me." Then they're like, "Is that how you feel?" And I was like, "I do. I feel like I'm being pet like a zoo animal." And they're like, "We're so sorry." But yeah. I had to put it on my desk.

—Sasha, IT Analyst at Fortune 500 Computer Software Company

Macro Hurts

I didn't know how to deal with the forms of microaggression . . . triple minority of being a woman and being Black and being young.

They were slow to give assignments, and they were slow to share their responsibilities. . .when you're the lowest on the totem pole, . . . if your team is getting the unsexy work, you're now getting the lowest of that.

—Kimberly, Manager of Credit Services Company

So, it's four of us: an African American female, who is myself, a Caucasian female, a Caucasian male, and this African American male. He speaks to these people on the telephone on a daily basis. They go to lunch. They've been to each other's houses. And they've never invited me.

—Maxine, VP at Financial Services Co.

It's me being too logical. But isn't it ironic that at the end of the day, they prefer being more comfortable than the bottom line? I mean, the whole perception is that White men command and control. All they care about is money. All they care about is the bottom line and making money and blah, blah, blah. But at the end of the day, if they know that you've got the goods and you could make their teams more efficient, a.k.a. make more money than this guy who doesn't know what he's going to do and could end up hurting our bottom line, it's going to take longer to get him up to speed, but they'll risk that just so they can feel more comfortable?

—Paulette, Senior Agile Coach

I often feel that women of color, when we lean in and we speak out and speak up, it's not taken the same if it came from a Caucasian woman, if you will. So, I don't necessarily feel that that advice is for us because I've learned in and I've spoken up and I stood up, and I've been sniped down, diminished, my voice taken away, my self-esteem destroyed, my lack of worth just totally stripped away from me. So, I don't necessarily feel that that advice was for women of color, because it's not acceptable.

—Laura, Senior Process Manager and Six Sigma Black Belt

I remember we were on [an] account last year. There were two of us deployed to an oil and gas company. I had one of my employees come into the room. She is a woman of color, and she was like—I felt like—she had just been introduced to the client, and she was like, "I don't know if he thought because I'm young or because of this that I didn't know what I was talking about, but I do. Why did he treat me this way?" She was very surprised.

And now, as a business owner, I have to look at that with clients. How are they going to treat my people because they may have some assumption based on how my people look on my team?

—Vivian, CEO of Agile Coaching Practice

[I was a] Marketing Manager, and there were rumors being spread that I was not qualified. So, I printed copies of my resume and sent them around the office to prove I was, in fact, qualified. As I look back on that situation long ago, it was a naive move, and I would not do that today. However, it did silence the rumors!

Speaking of approval of the marketing campaign, I was requested to be present at the review, which was a big deal at 25. So, I walked into the Executive suite and sat at the table with five Executives. One was a Black male executive with not much power. I remember the EVP looked at me, looked at the campaign presentation, and looked at me again. He paused and said, "Who helped you with this?" I responded, "No one," and he shook his head in disbelief. I truly think he believed I was lying. He requested that the campaign be reviewed by DDB Needham, an external media firm (one of the largest in the world). The report came back with zero errors and no recommendations. I held up my head and felt proud of myself. But, at the same moment, I felt defeated.

—Terry, Director of Strategic Transformation of Automation Company

One of the White guys went to his management team and to my management team and said that it was beneath him to work for me, and they did not move him to my team. So, out of that, I was like, "Wow, you get a choice? I didn't know there were choices in these scenarios."

—Jalessa, Process Manager at Financial Services Company

Then there are instances in which we are hired for our knowledge and experience, but we can't be the "face." We can't be the leader of the team, department, or organization, so another more acceptable person, usually male, is hired, and they are trained up to the work we already know how to do. And, of course, they are paid more!

I was hired as a Senior Scrum Master in anticipation for my partnership and assistance as Agile Coach for their transformation. I had only been employed for a short duration when they hired another Scrum Master who very quickly was accepted in the boys club. This person had no former experience as a Scrum Master and had lied on his resume. The company quickly aligned him with an Agile coach they hired as a consultant to get him up to speed. [He was later promoted to Manager and they made me report to him. . .enter a "pet to threat" situation of epic portions.]

—Terry, Scrum Master at B2B Automotive Services Company

So, I have years and years of experience. I have consulting experience, and [I'm going for] the VP of Agile Transformation. And I go in and I do all the interviews and the panels and the psychiatric tests, and obviously I scored high because it's between me and another candidate. The CIO of the company—I should have known something was going on. The CIO of the company calls me on a Saturday morning, and he says, "Oh... everybody loves you. This is great. We want to bring you on, but we have a really good program. It's between you and another candidate. And I know you wanted a VP title, but this is what we're going to do for you. We're going to give the VP role to this candidate, but we're going to give you a senior director role."

He lists all this stuff, and I'm really excited. I'm typical female, even more Black female, like, "Oh God, this guy is great. I could totally learn from him. This is a great opportunity. This is awesome." They're like, "We'll give you the same amount of money. You'll have the same amount of money."

. . . Doing my investigation, I'm like, "I don't know this dude, and nobody I know knows this dude." And then I see he's got, I think it's Brown undergrad, and then master's at Tech, and then PhD at Georgia Tech. Georgia Tech is really prestigious . . . .and I'm seeing that he was the director of programs at another company, and I'm thinking that's not really transformation. And I couldn't really see that he'd ever had anyone report to him. I'm like, "Okay, whatever."

You know us. Positive vibes. Positive. And I meet this guy, and kind of right way, I'm like, "I'm not sure if he really knows what he's talking about," to the point even our vendor [said something]. . . and they're doing modern Agile and they're well known. Even when we'd get into meetings with them, the leader of that company would always want to talk to me and not talk to my boss. And as I started to get through it, I'm like, "This guy knows nothing."

And as I started to finally do the connections and meet people in Agile who worked with him, I met a person through a friend. She was like, "Oh my God! Not that guy."

[Long story short, I left and without me, and he had to leave shortly after.]

—Paulette, Senior Agile Coach

Whenever I bring up some indiscretion that my peer had done, what I was always told, "Well, you're the bigger person. You have more empathy. He's X, Y, Z. You need to take on his bad behavior." Accept the apology when he called me "hun," right? Accept the apology and move on. It was almost as if, "You should be happy that you're here. You should be happy you're part of our inner circle. This is the role you have taken on, and I'm not going to let you have any other role."

And you know how we, especially as Black women, skirt around it. We keep professionalism.

You can't be angry. You can't be aggressive. Your cheeks hurt from smiling all day, that kind of thing.

—Paulette, Sr. Agile Coach

In general, I feel more of the bias for me being a Black woman more than me being a woman. I feel like a lot of times, when I walk into a situation, especially now that I'm among the manager ranks, that it's a shock factor no matter what I do.

—Diane, Sr. Business Consultant at Big Five Consulting Firm

I have been made to feel as if I am too much, I am too loud, I am too strong, I am too opinionated, I am too serious, I am too fierce. Even in some instances, I'm too understanding; I understand my people, versus, "Let's just get to the bottom line and hit this deadline."

—Stevie, Sr. Research Specialist at Multinational IT Services Firm

Interviewer: What would you say is the one most impactful situation, the one that stands out in your mind the most in terms of unfair treatment or discrimination or fill in the blank, whatever it might be?

Respondent: . . . Walking in the door every day.

—Charmaine, Senior Government Aeronautical Staff Member

As I indicated previously, sometimes microaggressions aren't actually "micro" at all, like how the cut stings a bit more when sexual harassment occurs at the hands of an African American man. For Black women, attacks and aggression committed by another woman are equally as appalling. However, there were so many examples of those stories from those I interviewed who offered stories of White women committing acts of oppression that I had to add a separate section dedicated to it. I must emphasize that there is a great divide between African American and White women in the workplace. Authors Smith and Nkomo, in *Our Separate Ways: Black and White Women and the Struggle for Professional Identity*, offer details of how the chasm was created; suffice it to say that gender does not always unite, and in the United States of America, race and ethnicity continue to define and divide us.

> [I] love him to pieces. As a manager, not so much. So, something that currently is making my blood boil and what I'm dealing with is he hired his grandkids' babysitter to do a job that we do that is a very high-profile job that most of us have had at least 15 to 20 years of experience. And now she's in a role where we work with the legal department. We work with high-end things for our company. It's a very big deal. Well, now, again, no shame how you got the job. I can even get past that. But her current way of being, she's extremely racist. Most of the team is Black. She speaks to everybody like they're something under her shoe. And she has since thrown things at me. She's come for me, and I really had to get real to the left with her. I let her know real, real quick I am not the one to be real, real ethnic and scared the bejesus out of her because that's the only thing she responds to. You can do anything, and my boss seems to see nothing. So, part of the reason why when I go into the office I try not to even be in the same space as her, two weeks ago, she threw a phone.

I was sitting next to a coworker. We were in a conference room. Me and another coworker, a Black guy, we were trying to untangle the cords. He's a bigger guy. I'm not so big. And he was trying to plug it in underneath, and his hands were too big. I said, "Well, let me do it because my hands are smaller than yours." So, I'm trying to plug it in. She decides—this White girl, she decides—to grab one of the tentacles, one of the little microphones, and try to untangle it.

. . . And she's pulling on it, and it slips out of my hand and my coworker's hand, the Black guy. So, my boss then says, "Hey, it's a new conference room with new equipment. Don't go breaking stuff." Well, the White girl says, "Oh," pointing to me, and said, "Oh, well, [she] is pulling it and yanking it." I looked at her and I spoke to her like she was five, like she was one of my nieces or nephews. I said, "Sweetheart, we're not doing this." I said, "I didn't pull anything. We're just trying to untangle it."

Well, she got angry because she looked stupid. So, while it was in her hand, she flung it at me. And my coworker on the side of me—because, again, he's a big guy—stopped it from hitting me in my face because I was sitting down. And I gave her this look like, "If we were not at work."

But I was so mortified, and we got through the meeting, and then she was trying to speak to me to fix it and clean it. And when it was time to leave, I just grabbed my things and [left].

—Regina, Senior IT Associate

Dents in the Ceiling 219

The director was present, and the other managers were present. I was one of six managers. Again, only Black person. I was not the only female, but I was the only Black person. And our director was a White woman. We went through the meeting, and I gave the report out on my team. And she had a number two in charge, who also was a White woman. So, as my peers are leaving her office, she and her number two stay in their places, which is the director at her desk, the number two in a leather armchair immediately next to her desk. And she says to me, "Stevie, can you wait for a second? Close the door."

So, I close the door, and she says to me, "I don't know if you know this. I noticed this. [Tammy] and I were talking about this. You seem so, I don't know, so serious all the time. Maybe you should get a little more peppy." And I was standing in front of the door. I remember this. *More peppy?* I walked over to the desk as close as I could physically get to her. And when I get really angry, I don't raise my voice. I actually go lower, and I said, "Let me be clear with the both of you. You do not pay me enough to be peppy. You pay me for results, and that's what you're getting here." And I turned and left.

Now, I don't know what would have come to my career in that environment, but for me, that peppy comment was the end. So, within a week, I was resigning from that position.

Why leave, you may be asking yourself? It was one comment. Inappropriate, bias, and poorly delivered, but still just one comment. Here's why.

In that particular instance, what I heard her say is, "While you're over here shucking this corn, I need you jiving as well. I need a happy Black person. I need you tap dancing and grinning and saying, 'Yes, sir.'" That's what hit me. And for me, it wasn't even whether it was her intention. We weren't going to have this big Kumbaya about, "Do White people understand the history of Black women?" Not necessary. I'm out.

—Stevie, Account Manager at Real Estate Firm

[This White woman called me colored], so we were riding along in the car going to visit the site, and I'm thinking, "Did I hear her say that?" because I was on a call, and I ended my call and listened a little bit more to the conversation. And they were using "colored," like, just as I'm saying "Black" or "African American," they were using "colored" instead. Once we got to our destination, I pulled my colleague aside and said, "Hey, listen. I don't know what's up with her, but we're going to have to have a conversation with her, because I know if one of my senior leaders comes and hears her saying that it's going to be a no-go because they're going to [say], 'These people are. . .they're ignorant, and that is not something that we're going to promote.'"

—Reagan, Mid-Career IT Consultant at a Small Boutique Firm

[I interview for this job that I went back to law school to get, and after being highly recommended for it by the hiring manager, a White male professor, I didn't get it, so the same position comes available again and I interview a second time.]

The second time, when I interview with her [White female who was the boss of the hiring manager], she tells me, "So, we know you very well. You've been at the company for a long time, gotten all your performance reviews, and there is no question that you could do this job. We are not concerned at all about the fact that you could do this job. But we are very concerned about your fit. In fact, from my vantage point, I think you are very "entitled." *Entitled?*

I said, "Entitled?" I said, "I don't know what I have said or done to make you think that. I come from humble means. . . .My first job, my science job . . . was more than I ever thought I would ever be doing in my life. Now I'm in law school, which of course wasn't part of any plan I had. My point is this is all icing on the cake. And, certainly, I don't feel like it's just going to be given to me. . . . all-day interview, two- or three-day interviews, I know that. But there is nothing that I know that I've said or done to make you feel like I'm entitled."

She was so irritated, like, I'm just like, "I don't know where you're getting that from, where that comes from, or what I have said or done to you to make," but she made it very clear that she felt like I was entitled. She felt like I just thought that I should have that job, or it should just be mine, and she was wanting to make sure that that was not the case. And so, I didn't get hired. At that point. . .I had to make a move. If I was going to go transition from science into the legal world, I had to get hired somewhere. And so, this is now my second interview. . .where I've worked 13 years, and this woman is making very clear to me that ain't no way in hell I'm about to get [this position].

. . .Karma's a bitch, boy. She was there for years, but she ended up—now, this is what got back to me, and I've heard it from a couple people—that she ended up sending a reply-all email on some kind of craziness that absolutely wasn't supposed to be a reply all. And in my mind, it's along these same lines. It's some kind of diversity something. She's talking crazy in an email about something or somebody that she thought she was sending directly to somebody that ended up being reply all to some executive leadership group, and she got fired. Not fired blatantly. She retired, but she got pushed out. And I heard from plenty of folks over there that it was absolutely horrible. It completely exposed her, and she got let go.

—Mary, Senior Researcher at Fortune 500 Company

[One of the younger White female coworkers gets intoxicated and belligerent during an out-of-town business trip.] Eight coworkers, [all of us dressed in] company clothes, plus our host person is experiencing this. Some of the restaurant staff is experiencing all of this, and so I apologize to them. And she makes the comment, "You don't need to apologize for me." I said, "Oh, no, actually, I do. This is not appropriate behavior." . . . Now this restaurant is in a mall, not outside. We're outside of the restaurant but inside of the mall still. So, then this person just continues to just go off, and I said, "Hey, this is really inappropriate." And so, then she says, "Well, you need to stop acting like a bitch."

But I felt so many things at that moment, and some of it was a little bit of helplessness, because I'm like, "Man, I want to react to this, but I know I shouldn't react." And I'm glad I didn't. But what I did do was as soon as I got back, I was on the phone with HR filing a complaint, and I gave an account of everything that had happened and all the names of who was involved and all of that.

So that is all documented... and the thing is that person never once acknowledged or said anything about my apology. Nothing. In fact, she just acted as though nothing had ever really happened. . . . I was flabbergasted.

—Denise, Senior IT Process Manager at Fortune 500 Company

To prevent microaggressions from turning into macro hurts, you will need to implement three important clues—an internal locus of control, a crash course in chess, and a triad to guide you. Microaggressions are hard to overcome because, like the iceberg analogy, they're not the blatant discrimination you see at the top but the insidiousness of what is hidden underneath. All the unconscious bias, stereotypes, and acts of discrimination are symptoms of a much larger, more malevolent evil that is socialized systemic racism. Combatting the deep-seated biases that are microaggressions will require your mind, body, and soul to conquer it and conclude as victor.

Clues to Heal Marco Hurts

Power and Control

The jokes, insults, and subtle comments that indicate stereotypes about our hair, culture, ability to be on time, and clothing all add up, creating a heavy burden. This additional weight is really the agony we feel over a loss of power and control. Power and control in the workplace make your career or break it because those who have it have more ability to move up and around as needed. This means that the more powerful you are, the more political capital you have, the more decisions you can make without being challenged, and the less fighting you need to do to be heard, seen, listened to, and therefore get your work completed.

We are knowledge workers in tech and can very rarely accomplish anything alone. We must interact with team members who don't look like us, think like us, and oftentimes live like us. Crossing the divide of race, gender, and ethnicity can feel like a trip across a canyon every day. Since we, as African American women, typically have the least political capital on our teams, we must get creative when maneuvering to get our goals accomplished. To be successful at doing so, you will have to assess your power and control. Even though it doesn't feel like it, yes, you do have a choice! Where is your locus of control?

"Locus of control" is a psychological concept that refers to how strongly people believe they have control over the situations and experiences that affect their lives. I learned this concept in my undergrad Intro to

Psychology class. I remembered it after I received some constructive feedback and took a long, hard look at how I saw the world and how I interacted with my teams at work, which was negative. I was, without a doubt, a glass-half-empty person. Remembering that an external locus of control meant that I couldn't change my circumstances and that I had to deal with things as they happened was a very reactive way to live life. I was in victim mode, thinking life just happened to me and I was incapable of doing anything about it. Up until that point, I was still thinking like a child. My parents divorced, I moved away from my grandparents, and I attended a college that was not my first choice. I felt that I had to accept things as out of my control, and the best I could do was to just get through it. I retained the "just get through it" mentality.

However, the alternative is an internal locus of control, which meant that if I didn't like how my life was unfolding, I had the power to change it. I worked hard over the next four years to shift my focus. I fought through all the "you aren't smart enough," "you don't belong here," and "you won't make it" tape playing in my head as well as the battles with my own internal critic to get up and get moving and to keep believing that I could change my circumstances.

I share this story because the fight continues. Every threat of a reorganization, every change in strategic direction, every challenge to a decision I've made about my team or an initiative I've led gives me the same thoughts, but quieting them has become a bit easier. I repeat to myself that I am in control, my circumstances do not define me, I am smart, I have worked hard, I have learned a lot along the way, my decisions are solid, and no. . . I'm not crazy.

My job, what I am paid for by my company, is to be a change agent, so I'm always fighting against the "what we've always done" mentality. How successful I am really depends on how open the organization is to change. By being "the only," or at least among a few women of color in the organization, you are a change agent as well. Our very presence signifies a departure from what has always been done. Therefore, your presence (or change) will bring about resistance, and you must remember that you have a choice to maintain your power and preserve an internal locus of control. Expect resistance and prepare for it.

Chess, Not Checkers

Your next clue is to be clear on what type of game you are playing. The only way to win a game is to understand the rules and be clear on what winning means. I stated from the onset of this book that collectively "winning" for us means getting more than just one or two African American women to the CEO seat of U.S. Fortune 500 companies. If that is the reward, what is the game? In her book *Only One*, Monica Brown weaves stories throughout her book of being the only African American woman in leadership positions and how her strategic moves landed her in her first executive position. She often states that you are showing up to "play chess, not checkers."

For those of you not familiar with either, think of chess as a long strategic game of war, whereas checkers is more like a street fight. Both games have some similarities, such as a 64-square playing board and pieces that must be moved around on the board to win, but just like corporate America, the number of captures allowed and the number of ways that your chess pieces can move around the board vary greatly. In summary, when leaders compare chess to a career in corporate America, they are describing the fact that it is much more complicated, requires much more strategic thinking, and takes more time than a simple game of checkers.

At the same time that it took me to move from individual contributor to Senior Director, my sister-girl friend of mine resigned, started her own consulting business and is President and CEO of her Agile consulting practice with a staff of five and growing. Regardless of whether you decide to fight the good fight in your company or start your own, it's all a game. If you are serious about learning the organizational game of chess, then Harvey J. Coleman's book, *Empowering Yourself: The Organizational Game Revealed* is your playbook. He writes, "You can accomplish anything in your career as long as you understand and accept some major ground rules of the game. You see, your career, and life in general, is a part of the game, but the ingredient that makes it all work is people. Yes, that's right, the game is about people."[57]

I had the pleasure of sitting and listening at the feet of Mr. Harvey Coleman at a BDPA Conference in Atlanta. It was a pivotal day for me. I heard Mr. Melvin Greer, Chief Data Scientist of Intel, and right after Mr. Coleman; both gentlemen changed my perception of my career, my life, and what I wanted my legacy to be for my children.

Mr. Greer challenged how I viewed technology as not just what I did at work, but as something I should integrate into my life and expose my children to in terms of not just playing video games but showing them how to create the games. Why hide what I do at

---

[57] Harvey J. Coleman, *Empowering Yourself: The Organizational Game Revealed*, 2nd edition, AuthorHouse, 2010 pp. 5-6.

work from my kids? Instead, I realized I need to invite them in to why I enjoy my job so much.

Mr. Coleman taught me to manage my PIE—performance, image, and exposure—and woke me to the fact that performance is table stakes. It's my entry ticket but also my image and, most importantly, how much exposure I get to senior leadership is my passport to other positions, assignments, and promotions.

Tech, like real estate, sales, and HR, is very much a relationship-based industry. No one company holds all the keys to innovation, which is why Microsoft has purchased companies from GitHub to LinkedIn, including oodles of small gaming and virtual reality companies. All while Apple has acquired almost 10 companies just this year and counting. Similar to these companies, we need access to innovative ideas as well as additional capital to grow, expand, and create strategic alliances to move ahead. This is where you start to look at yourself and your career as a company. This next clue is about you becoming a CEO right now.

Your Personal Board of Directors

You are your own company—a CEO of a one-person company. You are visionary, manager, and employee. Just as with every other company, private or public, you have a board of directors. To take control of your career destiny, who is helping you? Who sits on your board? Is it time to make some re-assignments? If so, let's review your board positions and their profiles, based on how Carla Harris defines them in her book, *Expect to Win: 10 Proven Strategies for Thriving in the Workplace*.

Advisor: Webster defines an advisor as "one who informs or gives a recommendation."[58] Carla Harris defines an advisor as someone who answers specific questions.[59]

Mentor: Webster defines a mentor as "a trusted counselor or guide."[60] Carla Harris defines a mentor as a trusted confidante that offers specific and tailored career advice.[61]

Sponsor: Webster defines a sponsor as "one who presents a candidate, one who assumes responsibility for, or one who pays for something"[62] while Carla Harris describes a sponsor as someone who uses their internal political and social capital to move their career forward within an organization and sponsors impact pay, bonus, promotions, and retention.[63]

The role of each of these people is very important to your career trajectory. For example, the

---

[58] "Advisor," merriam-webster.com/dictionary/advisor.

[59] Carla Harris, *Expect to Win: 10 Proven Strategies for Thriving in the Workplace*, Avery, 2010, pp. 101-102.

[60] "Mentor," merriam-webster.com/dictionary/mentor.

[61] Carla Harris, *Expect to Win: 10 Proven Strategies for Thriving in the Workplace*, Avery, 2010, pp. 102.

[62] "Sponsor," merriam-webster.com/dictionary/sponsor.

[63] Carla Harris, *Expect to Win: 10 Proven Strategies for Thriving in the Workplace*, Avery, 2010, pp. 102.

advisor is someone who is knowledgeable about your industry or your specialty domain. When Vivian sought promotion, she was turned down the first time and given the feedback that she needed to be "known for something." Her company was seeking a specialist, but she was an IT generalist. As she picked up the culture competency clue and started to observe her surroundings, she realized that Agile was the wave of the future and she needed to learn it and become an expert in it. She reached out to the department's Agile coach and asked him to be her advisor. He taught her exactly what she needed to know about the basics of Agile processes and principles and allowed her to shadow him while he coached new teams. She didn't need to personally like him, become friends, or even share with him her career plans; she just needed his knowledge.

A mentor is a mutually beneficial, usually long-term relationship built on trust. The key word is trust. You trust this person to share the good, the bad, and the ugly with you. They know your failures, your wins, your fears, and your hopes and dreams. This person is wise, offers you beneficial counsel, and you should never leave their presence feeling worse than when you entered. A good mentor worth their salt will always pick you up and dust you off, encourage you, and give you the tools to put into your toolkit to aid you in your journey. A mentor can be within and outside of your organization depending on what your needs are at any given time. And, of course, you can have more than one mentor at a time.

Lastly, and arguably most importantly, is the sponsor. This person has the political and social capital to speak your name in a room that you are not in to have you considered for a promotion, a stretch opportunity, a special project—you name it. All too often we don't even know who our sponsors are, and worse yet, we confuse our mentor with our sponsor.

It was March 2010. I had just returned from maternity leave after having my daughter and was on the list to be reallocated right out the door. All non-senior level project managers (PM) were being cut, and I wasn't eligible yet for a promotion to senior PM. I suspected my fate, but having an internal locus of control, I took the bull by the horns and started to leverage my network to line up a new position. As fate would have it, I didn't need to after all. My sponsor approached me to ask me if I'd be interested in a position in his area. "Of course!" I replied. My sponsor quickly and quietly removed my name from the cut list, and without fanfare or an announcement, I just showed up on his organizational chart the next week. The only reason I know for certain that I was on the list and was removed is because a senior manager let it slip that she was surprised to see my name on the department roster, as she was sure she saw me on the exit list. I smiled at the thought of my sponsor doing a cut and paste of my name from one list to the other.

> Sponsorship is the main thing. But by the time you get to [an executive level, it gets much harder] ... I was a vice president when I left [a credit card services agency], and [not having a sponsor] was a barrier. I was having a challenge moving from vice president level to senior vice president level, and it wasn't because I wasn't smart. It wasn't because I didn't know my stuff, because you heard my story, and a lot of people know that I know my stuff. And so that next barrier is all about who you know and who likes you.
>
> —Kimberly, VP at a Credit Card Services Agency

I tell people all this focus on kids is great. We need to look out for the future. But it's the senior-level people who are having the hardest time because now you're not the cute intern anymore. [They need sponsorship more than ever.]

—Lisa, Cybersecurity Consultant

Now that we are clear on the three roles of your personal board of directors and have established that the sponsor is the most significant role, the secret to this clue is not mixing them up! Never confuse your advisor, mentor, and sponsor, as it can prove detrimental to your upward mobility. Monica, who is a head of engineering, was a bit irritated that her junior engineering team member, who was another African American woman, was constantly complaining and over-sharing details of her life. Whenever the young engineer encountered an issue at work, received constructive feedback, or needed to vent, she would find her way to Monica's office. Monica, quite frankly, didn't care to hear any of it. She kept thinking to herself, "This girl needs a mentor or counselor—someone other than me to confide in—and I want to help her, but not like this.

When Monica shared the situation with me, a light bulb went off. I exclaimed, "She thinks YOU are her mentor, and you just really want to be her sponsor!" Monica started laughing and replied, "You know, I think you're right!" If that discussion had not taken place between Monica and I, that poor junior engineer would have lost her sponsor, the one woman in the company who was willing to take a risk on her and help her get her next promotion.

Your sponsor doesn't need (and, in this case, doesn't want) to hear your bad or ugly. Just the good. They need to know how capable you are, trust that you are a solid performer who is willing to carry their name forward in the company as a legacy to their leadership skills and will be a walking testimony of their superb ability to identify and cultivate top talent. Everything else you need to discuss is for your mentor.

I am often asked when I give the advisor/mentor/sponsor presentation, "Can I ask for a sponsor?" The answer is: "It depends." It depends on your personality, comfort level, your relationship and history with the person, and many other variables. If you need a sponsor, I will tell you what I do. . . I share who I am, what my career goals are, and what I am doing to achieve my goals. It's a 50/50 chance the person will volunteer to help me reach my goals. If you really want them as a sponsor, cultivate the relationship and offer to help them; this isn't quid pro quo. A sponsor must volunteer to cash in their political chips to help you. You can't force it.

I have many other stories of how accidentally conflating advisor, mentor, and sponsor can get you in hot water. The key is being clear with who you are talking to and in what situation. It is possible to have one person be all three roles, but exercise caution and clarity.

At one time, I had a supervisor who was all three, but when I came to him with a problem that I needed to vent about, I would say, "Got a few minutes? I need you to put your mentor hat on and listen to this. . ." Another time, I needed coaching in understanding the difference between capital versus operational projects and asked him to "advise" me. Another time, he asked me to consider a position at his new company, and he wanted to "sponsor" me as a manager. It is a rare person who can successfully wear all three hats, so I recommend finding three separate people to fulfill these positions, and please do not be scared to seek out advisors, mentors, and sponsors who are a different gender, ethnicity, and culture than your own. After all, if we are "the only" in our companies, we must!

We can be sponsors too. We don't have to wait until we have reached a certain status in the company or have been promoted to manager to sponsor someone else. Below is an example of a young analyst in her first year acting as a sponsor for an African American woman who was an intern and later an African American man:

> [The intern] was not performing as highly as she had been expected to be, so based on the degree that she was studying in the school that she was at, they expected more from her than what she was actually giving.
>
> So, there were other students who were giving about the same amount of low effort or possibly less, and the conversation was—I noticed a slightly different conversation. Around the other students, it was more so, "Is this work the right fit? Did we misalign their major? Could it have been something with their manager?" Then with the other girl, who I was kind of supporting and trying to help, it was more so about, "Well, we gave her X, Y, and Z, and her manager talked to her," and pretty much all of the, "Well, she fell short on all of these things."
>
> "Does she want to be here? Is this something she really wants to do? Well, her parents own. . ." She had affluent parents in the Atlanta region, and

Dents in the Ceiling 235

they owned a store or a business. And then people thought, "Well, if her parents are doing that and she's interested in these other things, it may be that this isn't a good fit for her."

I'm like, "What does her family business have anything to do with it?"

So, the reason that I try and stay on these boards and these teams is because I know that other students who may not have had the connections that I was able to make or didn't go to an HBCU or have a very strong network within the company that they went to don't have that protection or that influence.

And there was another case of another African American student. . . . they were talking about not giving him a return offer. He completed his assignment. It wasn't up to the level that they expected, but that's because there was a mismatch in the major to the assignment. So, he was an engineer by trade, and they put him on a lot of very technical tools, risk assessment in the security operations center, building these maps and building all these other kinds of tools that he had had absolutely no experience . . .

The fact of the matter is he was able to figure it out, able to get some output, and able to, at the end of the day, check the box to say, "Yes, I completed my assignment." But people were questioning, "Well, it took him this long. He needed this much help. He needed all this coaching. I don't think it's a good fit. I don't think he should come back." And, once again, it took me being in the room, knowing his case, and having a relationship with everyone at the table to say, "Hey, remember, this is not his major. He doesn't have any background in this, and we're expecting him to compete with people getting master's degrees in security. He's a sophomore in undergrad."

—Tamara, Application Security Analyst

Active Ally Clues for Microaggression

You are Not Colorblind

Colorblind, you say? I thought being colorblind meant that you can't see the difference between green, red, and blue. Professing that you do not see the difference between Black people and White people if you live in the United States of America is just plain absurd and, quite frankly, takes things a bit too far.

There is no nice way to say this—if you are seeking to be an active ally, please just strike the phrase "colorblind" from your vocabulary. We know what you mean to say is that you try hard not to actively discriminate, that you are checking your socially programmed biases and stereotypes at the door and would never use the "n-word;" we get it and know what you mean. We do not need you to pretend that we are not African American, Black, or from some part of the African diaspora. You can see our natural hair, our thick lips, curvy hips, and big brown beautiful eyes; you just don't have to comment on them at work!

I have learned the hard way that the definition of "professionally appropriate" does not mean the same thing to everyone. So, to be clear, for you to avoid some microaggressions, the comments, questions, and general inquisitiveness about Black culture must wait until you are in a private one-on-one session and preferably with someone with whom you have built a relationship over time. Without the trust, you will look insincere at best and like a jerk at worse.

Remember my golf outing with John and Mark and the car ride back to the office? Mark hadn't built enough trust or social capital with me to cash in a question about a hip-hop song. The question wasn't appropriate. I was a 19-year-old Black female intern, and he was a middle-aged, White male manager who clearly in some way knew that the lyrics of O.P.P. were stretchy, otherwise he wouldn't have asked.

Think critically about your relationship with the African American woman before asking to touch her hair, make a comment about how well she speaks, or make a joke about her being late to a meeting. If you are reading this and honestly had zero idea that any of this behavior was inappropriate or viewed as offensive, believe me when I say that it is! Err on the side of caution and assume that the African American woman on your team values her privacy, is filtering what she says and does, and is constantly surveying the room for emotional safety.

Once you have established a real connection and the wall is lowered, commence carefully with playing "20 Questions." Questions are important, as the answers will give you more clues of how to support and be an active ally, but be patient and build a healthy level of professional intimacy first. We don't need or want you to go "blind;" we really need for you to see us and provide respect and equality.

Agile Thinking

I am an Agilest, so I can't let you move on without offering you a few tools from my Agile toolkit to help you become an active ally. Agile is a popular term in the field of information technology that leverages principles from Lean Manufacturing and Design Thinking. At one point, Toyota car manufacturing plants were failing miserably. They instituted Lean principles to improve.

For example, every line employee was empowered to stop the line if they saw a failure or defect. They created a "pull system" in which a predefined number of parts were manufactured at each step along the manufacturing line to avoid an accumulation of too many car doors, for example, but not enough hoods. In the early 2000s, Lean principles were applied to software development and an Agile framework was born. Scrum, Kanban, Extreme Programming, and other methodologies were popularized as well, but Agile principles, such as high team collaboration, inspect and adapt, embrace change, and iterative and incremental development, remain.

Wondering how agile thinking can help you become an active ally? Well, it's guaranteed to help if you collaborate with diverse team members, embrace change, take time to review what is working well and what is not working well, and and think, "What is one thing I can do tomorrow to help a diverse team member?" High collaboration in Agile means you talk to the person you are working with; no longer do we communicate through an email saying, "I'm done, not it" and run to the next project, which is exactly what developers, testers, and business analysts do to each other.

I'm a PM by trade. I saw the 3:00 p.m. email on Friday where the developer hit his deadline but left only two hours for the quality tester to run her test scripts and no one to ask questions when needed. Throwing work over the wall wastes a great deal of time. In Agile and Lean, we detest waste. It's a waste not to have a productive and effectual working relationship with all of your colleagues. Only socializing with the ones you are comfortable with does not create a culture of inclusiveness or belonging. Don't forget the Diversify Your Network clue you learned in Chapter One, and remember the importance of embracing diverse viewpoints, accepting that your reality is really your perception and that others experience life, the company, and even the meeting that you were both just in differently. Seeking these other perspectives will serve to make you a better leader.

My favorite Agile principle is inspecting and adapting, taking the time to review what you did and reflecting on whether it was the best course of action. For example, I was prepared to create a roadmap for my team for the upcoming year, and I took a moment to consider that there was a different approach—I could include them on the roadmap creation and get their ideas and feedback before confirming it, which worked, since additional input from diverse perspectives improves the output. Retrospectives mean to "look back upon." Take the time to think about how you show up to your team, and if there are aspects of your behavior that you should keep doing, start doing, or stop doing.

Agile isn't just for IT folks. Agile can readily be applied in areas outside of technology, such as marketing, HR, and even to your personal life. Agile expertise isn't needed to be impactful, and don't overanalyze it. Start by setting a goal to commit to be a better ally. Write down two to three things that you can do in the next two weeks to demonstrate your intent of becoming an active ally, such as setting up a virtual coffee with a diverse team member or apologizing whenever you accidentally cut someone off in a meeting, and practice these behaviors consistently for two weeks. At the end of the two weeks, take ten to fifteen minutes and reflect on how well you did—did you make a mistake or have a misstep? No worries. Agilests embrace failure as much as change. How can you commit to rectifying your mistake? If you did great and exceeded your goals, then keep going, allow your momentum to build, and keep adding active ally behaviors to your list and build your active ally journey map without too much angst.

Legacy Building

This final clue is kinda a biggie! Why did you embark on your active ally journey to start? What experiences did you have and what values do you hold? Think Simon Sinek's *Start With Why* and contemplate what it was about you that led you on the ally voyage. I suggest that you think about your legacy.

The executive coaches I interviewed, mentors I have, and active ally panels I've been a part of all center around common core principles, which includes legacy building. You aren't going to be an IT leader at a company forever. Eventually, you will retire, hopefully healthily enough to enjoy all the vacation and golf you can handle, if that's your thing. How do you want to be remembered? Will they remember you as a leader in the company's culture shift, rushing in a climate of inclusivity and change? If you are in a position of senior leadership, assess your current mentee group. Look for diverse protegés to be in the group and consider sponsoring them. If you don't think they are ready yet to be sponsored, give them the constructive feedback on what they need to get ready. Think of those you sponsor as your legacy-builders. These are individuals who will remember you nominating them for a special project, promotion, or increased responsibility and, in turn, will be indebted. Having open favors to cash in throughout your organization isn't a bad position to be in, and it leaves a wonderful heritage.

If you find yourself feeling overwhelmed—after all, issues like systemic racism are huge and far-reaching—focus on the impact you can have on the people around you. "The reality is that individual actions can go a long way to make change,"[64] Jonathan Kanter, director of Center for the Science of Social Connection, says. Kanter wrote about his findings on microaggressions and bias and offers advice on how those who seek to support others (also known as active allies) can do so by placing their focus on helping others achieve.

Others' successes becomes your success, which builds a fruitful reputation for years to come. All the clues that you have learned thus far do not have to be implemented all at once or one right after the other. Simply assess yourself, examine your own bias and awareness level of the racial issues in your part of the world, and embark on your expedition of becoming an active ally.

Unfortunately, as columnist Dexter Thomas discovered, the reality is that Al Harrison, Kevin Costner's character in the movie *Hidden Figures*, never really opened the door to let Katherine Johnson in NASA's Mission Control Room. According to the book by Margot Lee Shetterly, Johnson "sat tight in the office, watching the transmission on a television." So, you see, active allies have a great deal of room to charter a new path and rewrite history. You get an opportunity to do the right thing, and instead of keeping the African

---

[64] McKenna Princing, "What Microaggressions Are and How to Prevent Them," *Right as Rain by UW Medicine*, Sept. 3, 2019, rightasrain.uwmedicine.org/life/relationships/microaggressions.

American woman technologist sitting "tight at her desk," you can invite her into the boardroom.

Breakthrough Tools

- Microaggressions feel anything but micro. Over time, emotional tax can form and eventually lead to physical issues as well.

- Being yelled at and disrespected is unacceptable. The old-school "command and control" in IT is eroding as companies flatten their hierarchical structures and embrace more people-centric principles such as empathy, vulnerability, and inclusion.

- It is a "pick and choose your battle" environment. Correcting every incorrect assumption or back-handed comment would be exhausting. You are already on alert with your tone and word choice. Managing your energy is more important.

- We know that in America, race supersedes class and gender, so do not assume that being on a team of mostly women means that there are not inclusion issues.

Starting Points

Women: Assess your PIE—performance, image, and exposure—the way that Harvey Coleman instructs, then start a plan to address any deficiencies.

Allies: Select a book, such as *White Fragility, Uprooting Racism, Everyday White People Confront Racial and Social Injustice: 15 Stories,* or videos such as Netflix's *13th*, the TedX playlist *Talks to Help you Understand Racism,* and the popular YouTube video titled *Systemic Racism Explained* produced by act.tv.

Chapter 7

Put Me in Coach

"You wanna fly, you got to give up the shit that weighs you down."
—Toni Morrison
Song of Solomon

It's been 20 years since singer, songwriter, producer, and R&B icon Erykah Badu told us to pack light. As with many aspects of life, even packing requires skill, but in this case, it's unpacking what we have learned in order make room for the stuff we need to carry on our journey. We must learn to unpack the emotional bags we've picked up from our journey through corporate America by learning to navigate being "the only" to managing the way microaggressions affect us emotionally and mentally. Some of the coping mechanisms we employ must be unlearned, the most critical of which is the choice to be silent.

Silencing our pain did not originate from our own individual choice—it was a survival mechanism that was required by our foremothers. Our great-grandmothers, grandmothers, and many of our mothers never had the luxury of voicing their pain, for doing so would mean dire consequences, including demotion or loss of a job entirely, which meant loss of livelihood for our families and, in some cases, physical harm. African American women in the 1960s weren't too worried about burning bras as a symbol for equality. We were fighting for civil rights, a fight that could leave us facing a cross burning on our lawns or a brick through our window in the middle of the night. At the same time, we were also fighting to be hired, and the notion of a fair wage was a stretch when we really just needed a wage in general! The reality is that Black women have always worked and Black families have always been two-income households. For American Progress, Sarah Jane Glynn reports:

> "Mothers have dramatically increased their participation in paid labor over the past 40 years. . . . These shifts over time are substantial. . . but the changes have not been evenly felt by all groups. Significant differences have always existed in the family and labor-force experiences of White women when compared with women of color, particularly Black women. Women of color, and Black women especially, have always been more likely to work outside the home than White women. This difference in labor force participation is deeply rooted in U.S. history around race, gender, and work. Black

women were always expected to work, too often in undervalued jobs with low wages."[65]

Once our foremothers did make it to take on office jobs, the notion of "put your head down and work hard" was cemented as a survival mechanism. This survival mechanism was then passed on to the next generation of daughters.

After sharing a particularly negative and highly impactful discriminatory interview situation that left Myra without the position she applied for, she said she never shared it with her father, to whom she was very close growing up.

"That was like my first entry towards being a woman of color in STEM. And I never shared that with my father. Never told him about that."

Myra, who is a senior director of project management at an international research company, then went on to share her thoughts in preparing for her interview with me. She said, "Okay, let me jot stuff down. And then, unfortunately, I thought I didn't need to jot it down, but so many things came to mind. And now, as I speak with you, even more is coming to mind."

The same experience occurred with Reagan, VP of Alliances at an IT Consortium. She stated, "I have suppressed a lot over the course of my career, and as I talk about it and I was writing down some things to share, it started to flow back to my memory." McGee and Bentley's 2017 research findings reveal that structural racism, sexism, and race-gender bias were

---

[65] Sarah Jane Glynn, "Breadwinning Mothers Continue To Be the U.S. Norm," Center for American Progress, May 10, 2019, americanprogress.org/issues/women/reports/2019/05/10/469739/breadwinning-mothers-continue-u-s-norm/.

salient in women's STEM settings. These experiences were sources of strain, which women dealt with in ways that demonstrate how we armor and mask trauma.[66]

However, not everyone has the same journey. Mary, a research patent attorney, sailed through grad school without much incident of sexism or racism. After graduation she took a position with a company and had been there almost a decade before experiencing what she remembers as any real discrimination. Ironically, it was when she wanted to break the mold and move to a position that required more responsibility at a pace faster than what was predefined for her that she was hit with a tidal wave of racism.

Mary explains, "I moved around quite a bit but overall had an absolutely fantastic career, and many opportunities too—traveling the world, presenting research, generating data, just truly becoming a research scientist. . . .It was awesome. But the one situation that I have experienced in my life that was unquestionably a situation of—I wouldn't even say it was 'unconscious bias.' This was straight-up just prejudice, overt, outright bias. So, basically, I had worked 12 years before experiencing unquestionable racism in being turned down for a patent research role upon completing law school."

It is easy to get discouraged. Before you give up or throw in the towel, consult with someone over the pros and cons. It's always darkest before the dawn, but unlearning to "unstuff" your emotions is extremely important. An IT analyst at a Fortune 500 software company, who we'll call "Sasha," is still early in her

---

[66] Ebony O. McGee and Lydia Bentley, "The Troubled Success of Black Women in STEM," *Cognition and Instruction*, vol. 35, no. 4, 2017, doi. 10.1080/07370008.2017.1355211.

career and finds that "there's a lot more of us now, but... I remember I wanted to leave. I would have been so sad if I left because I love my job. I love my company. I love the team. I've actually been promoted three times in the last three years, and I love the team I'm on now. And I do great work. But [when] those [discriminatory] experiences happen to us. . . I don't think people really realize how much that affects us and how much you internalize that when you're just trying to do your job."

All those experiences from being "the only," hearing jokes at your race or genders' expense, being touched, or inappropriately propositioned for sex, all the way to having to constantly fight for the staff, support, authority, and help that you need just to get your job done every day, day in and day out, will take a toll. After all, playing superman all day and then changing back to Clark Kent isn't easy. Hence the need for a respite with a group of empathizing coaches.

There's nothing like sitting at the feet of five experts in the field of women's advancement and soaking up their wisdom. These are professional executive coaches who have built their practices around listening to women in corporate America and helping them navigate the rough terrain. The coaches I spoke with focus on holistic wellness and two serve as professional development coaches, but they all understand the unique aspects of the complex intersectionality that exists with not only being a woman at work but a Black woman at work. There is "a disturbing silence that exists when it comes to Black women talking about their pain, whether emotional or physical. Silence is one way Black women have adapted to living in a world of interlocking oppressions."[67]

The emotional tax that is levied shows up in unique and subtle ways, but the strain manifests in overt ways in our bodies. Sometimes when we go silent and stuff the feelings of our negative experiences inside, it is as if we are drowning ourselves. We must figuratively "desahogarse" ourselves, a Spanish term whose literal translation means "to undrown."

Marla Rubio-Teyolia, Executive Coach and CEO/Founder of leadership development practice Culture Shift Agency, hails from a Mexican family and is a first-generation Ivy League graduate who teaches women how to "desahogarse" themselves. Marla has been holding spaces for women for over 25 years. She first started working with domestic violence survivors and has since worked with women in the arts and now tech. Marla's first-hand knowledge of experiencing microaggressions and attacks on her culture became prevalent when she moved from the West Coast to the East Coast, working in New York in a research lab where she found herself to be "the only." (All parts of New York are not the cultural mecca the city prides itself in being, especially in upper Manhattan, where Marla worked 20 years ago.)

Over time and through working in various roles, Marla has answered her calling to help women of color who work in corporate America, primarily in the tech industry, to unmute themselves, center themselves, and walk in their power. When I asked Marla how many women in her practice experience microaggressions or work in a toxic culture, she said without hesitation,

---

[67] Ella Bell Smith and Stella M. Nkomo, *Our Separate Ways: Black and White Women and the Struggle for Professional Identity*, Harvard Business School Press, 2001, pp. 97.

"100%! Angel, 100% of the women I see are experiencing it." She sees her work as healing work because she views microaggressions as violence. They are daily attacks on your character and spirit, and that is violence that needs to be healed.

When Marla transitioned from her home culture to New York, shortly after sitting in the male-dominated research lab and baring the overall questioning of her peers at the Ivy League institution about her ability to perform, she started to wonder if she was slipping into depression. As with many of us, she pulled out the defense mechanism of armoring to protect herself. But Marla is concerned about whether you are you armoring correctly. There isn't anything wrong with protecting yourself, but you must learn to do it the right way, otherwise nothing gets in, but nothing gets out either.

Ninety percent of Marla's clients are high-performing corporate mavens who work in tech companies. She said within a year of these healthy women starting in a leadership role, they have high-blood pressure, are borderline pre-diabetic, and suffer from high cholesterol, all due to chronic levels of stress. Marla teaches that heightened levels of stress lead to high levels of cortisol. Cortisol, the primary stress hormone, increases sugars (glucose) in the bloodstream, enhancing your brain's use of glucose. Too much stress hormone can lead to weight gain, especially around the back and midsection, acne, headaches, irritability, and a host of over negative health consequences. Operating from a constant state of stress dramatically decreases your ability to think creatively, be innovative, and access executive-level thinking.

Marla stated that a lady in her practice recently told her that George Floyd was a horrific, violent, and awful event, but that the Amy Cooper video was what took her "over the edge." Amy Cooper was the woman who called 911 on a Black man and lied and said he was threatening her in New York's Central Park. Marla's client was so upset by the viral video because she said that she "works with an Amy every day," a White woman who weaponizes her White tears and becomes fragile in the presence of a Black person, purporting herself as a victim. This was the daily experience for Marla's client.

Regarding the massive amount of stress, Marla says, "This is all very, very real." It is important for us to examine all the various ways in which we have been indoctrinated as a society and culture into a system that benefits from our silence. We were already full humans to begin with, whether White men and women in positions of power see it or not. While voicing our experiences and bringing a light to our pain is the main purpose of this book, it is not enough to just give voice to the pain without working to shed the excess baggage and help you reclaim your identity and your gifts that may have been unintentionally tucked away.

Marla suggests that we hold space for each other, heal ourselves, replenish ourselves, disrupt our negative mental thought processes, learn to properly breathe, and then be ready to create a strategy. "All that healing and feeling good stuff is great but doesn't mean anything without a strategy—without putting a plan into execution," Marla says. Holding space for each other looks like a virtual Zoom session with the girls. I call mine "Wine Down" time.

One night, myself and three of my sorority sisters grabbed our wine, moscato (and for one, a bottle of rum), to laugh, talk, complain, and release. One soror was dealing with an Amy Cooper, another a Karen, and another, since she is a healthcare worker, was running on fumes from dealing with multiple health disasters! After our Friday night 90-minute therapy session, I had never slept better. I woke up at 5:30 a.m. Saturday morning, refreshed and in a great mood.

While you are holding space, be on the lookout for negative thought patterns in yourself and your friends. During my Wine Down call, whenever one of us started down the "I don't know what I'm going to do" path, the rest of us rushed in with "you got this," "you are an amazing person and worker," or "don't listen to the enemy, you will get through this as you always have; you are a winner." We all need a verbal disruption from true friends or mentors to prevent negative mental spiraling.

Iyanla Vanzant conducts virtual spiritual spas. When I attended my first one, she shared ideas such as creating an altar for yourself—a sanctuary to go to—that includes your favorite items, candles, soft music, and pictures of your strong women ancestors to pull strength from. Use all of this to center yourself.

Finally, Marla reminds us to breathe. Just breathe. She has studied many different healing and breathing techniques from a shaman that she shares with her coaching clients. During our interview, she taught me two three-to-five-minute exercises that help slow down your heart and move your breath lower from your throat and chest into your diaphragm, as well as how to perform a healing self-touch to decrease some of those cortisol levels. She teaches that we have a "wealth of free wellness resources inside of us and that wellness doesn't equal Whiteness." Too many yoga pants ads, yoga classes, and detoxifying eating plans are marketed toward White people in general and White women specifically. We can find joy and healing in those restorative practices as well. If we don't feel comfortable being "the only" in yet another space, like a gym or yoga studio, perhaps be like Renee and start your own!

Dents in the Ceiling 253

It was three to four years after having my daughter, and all my exercise routines had become non-existent. I was under pressure to make Senior Program Manager and with that came a lot of late nights, pressure to over-perform, and very unhealthy eating habits. I no longer would eat at the café; I needed each and every minute during the day. I worked vigorously for meetings or to access resources such as the plotter machine to print large network diagrams that I would roll up and pack to race out of the office right at 5:05 p.m. to beat traffic to pick my daughter up by 6:00 p.m. sharp! Every working parent whose child is in daycare knows the 6:01 p.m. charge—the $5 fee that hits your account for every minute you're late for pick-up. A fee that must be paid to return your child to daycare the next day. Well, I hate that fee. Since highway construction and traffic is unpredictable, you never know when that random car accident can cause a backup, so you better build in buffer travel time to avoid the dreaded daycare late fees.

Well, this early to rise, late to bed, go, go, go routine left little room for exercise and healthy eating. I kept telling myself that this was temporary, just until I made Senior Program Manager, and then I'd slow down. But when I made Senior Program Manager and had four program managers reporting to me, I didn't slow down. I had become accustomed to the pace and so did my body. I craved the adrenaline. I couldn't sit still.

I then noticed a few of my peers coming to work with large containers of water and carrots and celery sticks for snacks. They looked great and all had lost weight. Enter Fit with Renee. Renee is a certified personal trainer, former bodybuilder, and health and fitness expert. An African American mother licensed in dentistry, she, too, looked up one day unsatisfied with her weight and embarked on a personal fitness journey that led her to become one of the most successful fitness instructors in the Midwest.

I can't imagine how many women she has helped, including myself. The cool part is that 99% of her practice is made up of African American women. Doctors, lawyers, engineers, HR specialists, VPs of finance departments, you name it, are all coming to Renee to lose weight and gain a healthy relationship with food. However, what they found was much more than that.

What I found was permission to take care of me. I prided myself on taking care of my family and giving to my daughter. Any time away that wasn't for work or community service, I felt guilty. The first step—and for me, the most challenging—was shedding the lie that I could give and give without filling up my cup first. One of my favorite project management gurus, Laura Barnard, always says, "First apply your oxygen mask before helping others." I lost twelve pounds, learned clean eating techniques, such as always trying to pair a carbohydrate with a protein (which I still struggle with today), but most importantly, I never went back to the fallacy that I don't have time for me. I'm now four years after baby number two and back working out with Renee.

Creating and forming healthy habits is the key to creating a holistic wellness plan for yourself. Dr. Dahlia Henry-Tett, a certified Health Education Specialist with over 20 years of experience, is an award-winning professor in the Health and Physical Education Department at a college in Virginia. She is a health educator and wellness consultant who focuses on teaching healthy habit plans. Dr. Henry-Tett says that it is important to set our intentions. We must be intentional about self-care and with that comes giving ourselves permission to practice self-compassion. Once our intention is set, we can practice healthy habits "SMARTLY," which means:

> **S**elf-Care
> **M**indfulness and Gratitude
> **A**ctivities you Enjoy and Exercise
> **R**outines and Consistency
> **T**alking and Self-Talk
> **L**ens/Perception
> **Y**ou owe it to yourself and your loved ones to manage your stress and health effectively.

Dr. Henry-Tett encourages us to leave a legacy of wealth for our children, and she defines wealth as modeling strong faith, resilience, and demonstrating mental wellness. Senior Agile Coach Paulette recounts a time in which she resigned after a long and nasty battle with her White female supervisor, in which they fought over decision-making of a strategic transformation. "I remember going back, and it was *The Twilight Zone*. . . . She didn't even announce I was leaving. I left, and then finally—people were pressuring her so much because I had done so much work in the six months I was there. She kind of snapped and said, "Look, Paulette left. She's not here anymore, and I don't want to talk about it anymore." When I said I was traumatized from that experience, I took four months off last year."

Paulette needed four months before making her next career move. Her personal situation afforded her the much-needed time to re-center herself, recharge, and reconnect with her spirit. When the notion of finding yourself starts to emerge, that means that somewhere along the way, you must have gotten lost. If you need permission to take some time to go find yourself, look no further than well-being coach Radiah Rhodes.

Radiah Rhodes is a wellness innovator. She is a visionary, entrepreneur, designer, engineer, and now executive coach. After two decades of experience leading in small businesses, Fortune 100 companies in the beauty and information technology industries, she joined with two other powerhouses to create Evók.

Radiah says she sees far too many of her clients using escapism to cope, increasing their wine consumption to help them relax at the end of a day filled with tension. Her practice is filled with very ambitious women (women wanting to write a book, start their own business, run a department) who were getting the message that they couldn't do what they wanted. This led to a decrease in their self-esteem, confidence, and even a drop in their overall credibility. While verbally being told you "aren't a good fit" might be very subjective feedback, it leaves the recipient with no place to go for improvement.

Instead, Radiah educates her clients on the need for holistic wellness, which includes financial and relationship wellness as well as physical wellness. As the author of *Being is the New Doing: A Divine Guide to Owning Your Energy, Time and Peace of Mind*, Radiah encourages us to start with our "who." Who is about identity versus why, which is really a catalyst for doing; before we get to doing, we have to go back and do some work to find out who we are first. Simon Sinek's bestseller, *Start With Why: How Great Leaders Inspire Everyone to Take Action*, started a whole "why?" movement for corporations and executives in the

United States and abroad. However, Radiah challenges that assumption. She says, "White men can 'start with why' because they are known; we have to deconstruct the 'who' that the world defined for us, because a Black woman's identity in American society is too limiting."

Evók leaders take their clients through a nine-step internal journey that helps to build the muscle of agility and power based on who it is the client wants to become and how best to define and then reach her goals. Radiah says that there is a great deal of time spent on the three S's: Struggle, Suffering, and Sacrifice, and then learning how to release the inner energy to propel you forward. It becomes unproductive to sit in negative energy—you absorb it. We must watch the environment that we allow ourselves to sit in and be mindful about how it is impacting our inner beings.

Evók also conducts an assessment on intentionality for companies. Companies can score from negative nine to neutral to positive nine. Most of the companies that Radiah has experienced score negatively with the best being neutral. What are the intentions that are being scored? Their intentions on how they treat Black women in their workplace. Similar to the Althea Test founded by Anjuan Simmons, if you can assess a company's intention to their underrepresented workforce by examining their DEI practices, hiring and promotion outcomes, and general culture of inclusiveness and belonging, you get a pretty accurate idea of how welcoming or toxic the environment is.

This translates into how healthy or unhealthy the workplace will be for us as Black women. The women who run Evók know that working hard isn't the answer. It isn't serving us as a whole, especially the many Evók clients who are near burnout from overworking themselves. Radiah and the ladies at Evók help their clients build themselves from the inner person out so their clients can show up to networking events and virtual Meetups attracting the right people into their space which can result in a positive win-win relationship for everyone.

Crystal Kahlil, author of *Hard Workers Work Hard, and Networkers Move Up!: Accelerate Your Career 10x Faster*, also ascribes to the notion that "putting your head down" and remaining silent while you sit in your pain is not the answer. Khalil recommends assembling your tribe, creating a personal board of directors, and being expansive in your network to help ease the misconception that African American women need to "just work harder to move up." Khalil emphasizes in her book the need to create a very strong network. These networks can be successful in serving you, but you must show confidence and show up powerful for others so they can see that connecting with you will be beneficial for them. But what if you don't feel powerful yet?

Sally Helgesen, international speaker, leadership coach, and author of the bestseller *How Women Rise*, suggests that we borrow some power until we can self-generate it. Accomplishing that means that we must get comfortable asking for help.

Sally explains that generating power can come through our relationships with our best friends, our tribe, our posse, our girl group who will always have our back. The tribe should serve two main purposes:

1. a place to vent and let our hair down to be us.

2. a place of positive outcomes focused on solutions.

Dents in the Ceiling 259

Sally warns that girl groups can get negative quickly and turn into "gripe fests," and if that happens, find a new group! Just like my "Wine Down" Zoom sessions, after we tell each other our work and family woes, it is always followed by uplifting positive affirmations and sharing how much we love each other and the best thing about each one of our personalities; we end with providing each other positive suggestions on how to get through our prospective storms. Having sisters on speed-dial during a work crisis to remind me to breathe or offer a much more practical—less emotional—perspective is often just what I need to shift my perspective and plug back into my internal locus of control and confidence.

Sally is a strong proponent of having a personal board of directors as well. She cited Carla Harris as a wonderful example of an African American woman working on Wall Street who took some "tough" feedback but worked to build her network of allies, which included advisors, mentors, and sponsors to help her. Sally reminds us that having a network of people with different experiences is a must—for example, they don't have to be in the same profession, and we must include men in our ally groups. In Sally's experience, people who have successfully navigated major career moves never did it alone. The number one common factor was that someone was helping them. She emphasized, DO NOT try to do it alone! Allies can be used to:

1. Set our intentions. Why do we want to be VP, lead an engineering department, or make our way into whatever role we are striving for next?

2. Build our visibility.

3. Help in a crisis.

The confidence to focus on who you are is exactly what our next executive coach, Sherrie Brown Littlejohn, says her clients need. Sherrie Brown Littlejohn is the definition of a powerhouse executive. When she showed up to our EMERGE fireside chat session one evening, I was riveted. The gems of advice she was giving were dripping through my laptop monitor like pearls and rubies. Sherrie speaks with a level of confidence and no-nonsense candor that is unmistakable. She left such an impression on me that night that I reached out to her before the call concluded to ask for an interview, and she immediately accepted.

Sherrie Brown Littlejohn certainly knows about navigating the corporate American ladder to the top. She is a business veteran who started in Bell Labs then worked at Pacific Bell. She is armed with a background in architecture, design, and successfully executing on strategic initiatives. She rose to become EVP-Head of Enterprise Architecture and IT Strategy at Wells Fargo. Needless to say, Sherrie has a wealth of experience and achievements under her belt that she now shares with her coaching clientele, who are the next generation of high-powered, driven, extremely intelligent African American scientists and engineers who still struggle to be heard and have their accomplishments recognized.

She started her coaching practice, Littlejohn Leadership Coaching and Consulting, along with a blog to do what she's always done, which is to coach and mentor those who need her help.

Sherrie said that what she is seeing in her practice is that crisis moments are the cherry on the top of the stress cake that we feed from daily. She said her clients are strained just trying to find time to think!

Mothers experience increased responsibilities, such as serving three meals a day, caring for kids, setting up virtual homeschooling, all-day Zoom meetings, which in some organizations means being camera-ready; then, the day is gone, and Sherrie's clients said they have no time for themselves! Landing the next promotion, networking virtually or otherwise, and practicing self-care are taking a backseat to just getting through the day.

This situation sounds eerily familiar to the one I described earlier when I was going for a senior program manager role as a new mom. You may say, wait, Angel. . . your daughter was four at that time. You weren't a new mom. I challenge that—when you are the parent of one, you are always new, as each stage comes with behavior, challenges, and victories that you've never experienced before. So, new mom or mom of multiple, or mom of fur babies, times of severe crisis add some new stressors to the plate, and if you are "the only," experiences of microaggressions, sexual harassment, or being bullied before a crisis hits, I guarantee you that you need some new coping skills.

Sherrie highly recommends meditation. Meditation during your morning routine can help start you on the right track, center your inner being, and help set your intentions for the day. There are many types of morning routines to select from, and the key is experimenting with a few—use an Agile mindset and try one for two weeks, then do a personal retrospective on what you did or didn't like about it. Then try another routine and blend methods of more than one to tailor to your needs.

Kailei Carr, Beyond the Business Suit founder, women's advancement expert, podcaster, coach, and mentor, introduced me to two different morning routines during my time in the EMERGE Leadership Program. They are *The Miracle Morning* by Hal Elrod as well as Kailei's own morning meditation she calls CLAIM Your Day. I found benefits in both and tweaked them to fit my life.

I've found that we need different routines for different stages in our lives. I look forward to the day when both kids are dressing themselves in the morning and cooking their own breakfasts to give mommy a few extra minutes in her morning routine. Until then, I cherish my 15 minutes of "me time" each morning, and without it, a very different Angel emerges from the bedroom.

Just as important as meditating in the morning, Sherrie also encourages us to get grounded in *who* we are, just as Radiah coached. This means that we must work to define our identity, our goals and desires, and become comfortable in our own bodies, all of which will position us to increase our confidence. It is important that we don't play the victim role and get into the mindset of what is being done to us but instead focus on the fact that we are doing the best we can with the tools we have been given. If you need more tools, different tools, or a whole new toolkit, then set out to do the work to make the necessary changes. While you're working on a personal improvement plan, creating that Plan B, or studying for another degree, don't forget to take the time to celebrate who you are and love yourself.

Sherrie said, "Celebrate the fact that if you are in STEM and getting a paycheck, you have already made it!" You don't have to reach a certain title or position to claim success; you've already beaten out plenty of other qualified candidates for the role you are in now because you are the best person for that position. If you don't like the position you have now, you have every right to work to change it. While you're in the role, however, there are some improvements you can make to help the person after you.

A part of being grounded means that you seek out others' opinions to add to your own. You have a perspective and a viewpoint but incorporating other's views is an important aspect of a great leader and will help you learn yourself as well. Sherrie encourages us to "become curious" but drop the "whys." Don't use the word "why;" instead ask, "Help me understand?" Or, "How could I have done that differently?" Or, "How do you think that situation occurred?" Getting curious is an open and collaborative stance. It doesn't negate your position or viewpoint, but it allows for others to be let in—a phenomenal step in creating a permeable armor!

Sherrie finally warns us to guard our mindsets. Too many of us think we must have all the answers. Sherrie reminds us that by the time you make it into a senior leadership role, like a director, senior director, or above, the expectation of your contribution changes. No one expects you to have all the answers, but your job now becomes orchestrator, bridge builder, visionary, guide, great delegator, and being skillful at synthesizing diverse ideas and opinions.

Her advice reminds me of the book *What Got You Here, Won't Get You There* by Marshall Goldsmith and Mark Reiter, in which the authors explain that the skills at your current level are not necessarily the same skills that you will be using at your next level. Oftentimes, White men are promoted on potential while minorities are promoted on performance. Therefore, we must learn new skills in our current roles and demonstrate them now to be considered for the next job up.

Sherrie gave an example of how one of her clients was worried about how to write up her pitch to become a director. Sherrie told her, "You already are a director, and you've been doing director level work for a year now, so the promotion is just a rubber stamp, a signature." Once you see yourself as a director and have that confidence, others will as well.

For example, I was PMO Manager but performing director-level work. My supervisor was the PMO Director, performing VP-level work. Despite our titles, we both showed up as Director and VP in the spaces we operated. We didn't give our titles; we just acted as Director and VP to the point in which others just assumed those were our titles. When I showed up at the IT Director's Council, did a bit of imposter syndrome kick in? Just a bit, but I was confident in my job duties. I knew I was bringing a director's mindset to the table, and I didn't have to wait too long because the title eventually did catch up to reflect the work that I was already performing—just as Sherrie indicated.

When going for that next step on which you've already been working, be sure to phrase your accomplishments with "I" first, then "we." For example, "I reached out to marketing to help dress up our customer pitch deck and together we created a new template that the customer loved." So, "I guided," "I recommended," "I suggested," "I strategized with Sue, Bill, Jim, Frank," "and then we. . ." Your achievements are always "I" then "we" to emphasize your role in the collective win.

Not all your counsel needs to come from women of color. There are some well-intentioned and passionate White men who are advocates for women who offer executive coaching, mentoring, and sponsorship willingly. Jeffrey Tobias Halter is the former Coca-Cola Director of Diversity Strategy, TEDx speaker, author of *WHY WOMEN*, and advocate for women's advancement and leadership in the workplace. Jeffery has dedicated his work to ensuring that White men see the need to sponsor women, and he also shows them how.

Brandishing a pair of red pumps on stage for the first half of his 2018 TEDx talk, Jeffrey literally demonstrated that it isn't easy to walk a mile in our shoes. As the president of the gender consultancy group YWomen, Jeffrey says that he didn't choose to become an ally at the onset. He was placed in his role on the D&I leadership team with a bit of a head scratching as he thought, "Why me?" He said that he had his "White male epiphany," also known as the acknowledgement of White privilege and that the world revolves around White men, challenged after he decided to get curious.

Just as we heard from Sherri Brown Littlejohn about the hallmark of great leaders being getting curious, Jeffrey Halter went about his network having open and honest conversations with the Black members of his organization. Hearing about racial profiling, microaggressions, and now in his role as D&I leader, he was privy to the dreadful statistics of hiring and promotion of women and minorities at his company. This made him think that since White men made up about 80% of the leadership positions, then instead of being 80% the problem, they could become 80% of the solution.

Jeffrey's thinking was correct. His company embarked on a journey to eradicate systemic and institutionalized racism, by targeting programs toward White men. His company received the Catalyst award for their D&I efforts. Although they never successfully got an African American woman to president of a division, they made great strides in breaking down the barriers that it would take for her to get there under his watch. After his D&I work in corporate America, he embarked on launching YWomen.

There were some barriers like the ones that we've already read, and Jeffrey admits quite frankly that grown men can act like frat boys at times. He retold a story, although not proudly, in which he and a group of coworkers purposely changed the team meeting time to 7:00 a.m. from their 9:00 a.m. usual time to haze the new woman from corporate coming to their team. They wanted her to learn the culture she was coming into, which was hard, as most of the crew had to get up well before dawn to make their runs. It was an early shift work culture, and she needed to learn that if she was going to be accepted.

After about two months, she finally called out the absurdity of the early morning team meeting and changed it to 10:00 a.m. The guys weren't hazing her because she was Latina or a woman; it was because she was from the corporate office. Of course, she didn't see it that way; from her perspective she was jumping through all the typical mental hoops wondering, "Is it because I'm brown, or it is because I'm a woman, or both?" After Jeffrey had his epiphany, he admitted to her what they had done to haze her, and both can now laugh about it later and recount the lessons they learned from it.

I had a friend tell me that just after she graduated with her Engineering Degree from Purdue University, she worked the manufacturing floor at a local company where she was "the only" and the youngest. One day, she returned to her desk from validating the equipment and went to put her clipboard in her desk, only to find a frog in her drawer. Since she grew up with all brothers, a little frog wasn't going to scare her, so she sighed, picked it up, and let it loose at

Dents in the Ceiling 267

the door. She returned to her desk in silence while she walked past the snickering of her male colleagues.

    Just as Monica Brown, Harvey Coleman, and many others use chess as the analogous game to navigate corporate America, Jeffrey likens it more to *Game of Thrones*, the medieval fantasy drama. Like on the show, alliances are formed, truces are made, revenge fights occur, backstabbing is normal, and the typical executive's career resembles the climb of a lattice instead of a straight ladder. Jeffrey admits that in this corporate culture White men cover or hide their shortcomings too. In fact, according to Jefferey, "40% of White men cover as a way of playing the game and come up with strategies to overcome it."

    So, Jeffrey offers a few strategies we can use to help improve our odds of winning, starting with the Rule of Three. He cautions that women tend to stay in their current roles for too long. He said, "You don't have to master a role before leaving it." This statement gave me pause, as it certainly went against everything I had known and believed, which was that you must master a job before even being considered to move on to the next.

    Jeffrey's magic Rule of Three states that in your first year of any role, you are learning, the second year in the role, you become proficient, and in the third year of your role, you are most likely ready to move, as the incremental gain in knowledge you receive will start to hit a level of diminishing returns. This means that the more effort you put into it, the less and less benefit you will receive from it. Next, women are in a double bind that we will have to push our ways out of to be successful. Jeffrey said, "Men are scared to death to give feedback to a woman of color because they are scared of saying something wrong." The issue is that we need that feedback to grow, hence the binding situation.

The "angry Black woman" stereotype is real and rears its ugly head during performance discussions and when it's time to offer constructive feedback. Confrontation is the best way to deal with the situation, but be patient with your White male supervisor to allow him to see that you aren't going to lash out. Keep working on the communication until emotional safety builds in both parties. Remember, regardless of whether you think the feedback is BS or not, it's coming from a place of bias versus authenticity, the feedback is a gift, one that you can choose to receive or not. If you do get BS feedback or highly subjective responses, Jeffery encourages you to ask for data and dig for the answers you need.

Finally, sponsorship is key! Just as our friend and gender advocate Sylvia Ann Hewlett tells us, it is all about having a good sponsor. The White men that have made it to the C-Suite have had great sponsors to help pull them along, and guess what? The Black women that have made it to the C-Suite have had great sponsors to pull them along as well. Everyone who got there had a sponsor.

However, Jeffrey sadly admits that many Black women he has personally seen that have landed executive level roles were brought in from the outside. Many companies like to say that they "grow their own,'" but that is rarely true for the Black women among the ranks. This means that a significant advance will most likely lead to leaving your existing organization to make a leap forward.

For those who are younger in their career, 25 to 35 years old, Jeffrey says to consider working for a start-up. We are often so worried about paying back student loans or repaying our parents for college money we borrowed that the thought of working for an organization that isn't stable and barely making payroll is a turn off. However, if you can endure a few more years of macaroni and cheese in a box and a few more nights of cereal for dinner—like we did in grad school (not sure about you, but my fellowship stipend was funding my survival only), then the exposure to supply chain, R&D, sales, and product management will position you for manager roles in those departments. Your climb through those types of organizations rather than the supporting functions, such as HR, communications, marketing, and IT, will serve you better. Again, nothing is wrong with those roles, but rarely is the president of marketing made CEO. As we heard before, owning a P&L and managing the company's core functions versus the support ones will better position us for the C-Suite.

A study conducted by Hall, Everette, and Hamilton-Mason found how workplace stressors affect the lives of Black women and, more importantly, how we cope with the stress. Five basic themes emerged from their study when racism and sexism are experienced as stressors for African American women in the workplace. The themes are: (1) being hired or promoted in the workplace, (2) defending one's race and lack of mentorship, (3) shifting or code-switching to overcome barriers to employment, (4) coping with racism and discrimination, and (5) being isolated and/or excluded.[68] Their research greatly mirrors the findings from the 30 women who shared their work

---

[68] J. Camille Hall, et. al., "Black Women Talk About Workplace Stress and How They Cope," *Journal of Black Studies*, vol. 43, no. 2, 2011, pp. 207-226.

experiences with me and solidifies the need for us to take seriously the stress to our minds, bodies, and spirits.

We must act now before we buckle under the weight of it all. I encourage you to seek out an executive coach for yourself and put it on your development plan. Perhaps your company will pay for all or a portion of your coaching instead of sending you to a several-thousand-dollar class or one more certification that you don't need. Think critically about how a few more letters behind your name will serve you, because having a coach like Sherri, Marla, Radiah, Kailei, or Jeffrey may yield a greater return.

Chapter 8

Resiliency Building

"If you are silent about your pain, they can kill you and say you enjoyed it."

—Zora Neale Hurston

Imagine a Black woman sitting in a C-Suite-level boardroom. She is dressed professionally in a business suit on one side of the table. On the other side of the table is a collection of various men, White men, C-level executives in understated black and blue suits and matching ties. Now ask yourself, why is she there? Is she being interviewed for a promotion? Is she being reprimanded for her attitude, for being too aggressive when she thought she was being as assertive as her male colleagues? Is she there to formally begin the proceedings of her sexual harassment or workplace discrimination claim? Is she there as a consultant to give a presentation, and she's mistaken for the wait staff and reminded to make sure there's enough cookies and water for the afternoon meeting?

No matter how you've answered the question as to why the Black woman is sitting before a board of White men, you can imagine yourself in her position and recall the responses you received to your presence when it was your turn before a board of men who had the power to decide the future of your career. The questioning of your presence and the assumptions associated with your position in a place where you are thought not to belong will most likely keep happening for some time to come. Systemic racism and misogynistic attitudes will not be exterminated overnight. As a result of George Floyd's murder, most corporate organizations are just waking up to their internal racist practices and are now trying to "stand" with us, but while they are standing, while they get their bearings and learn to crawl, we are still here and still fighting because we are resilient.

Resilience in the corporate structure requires that we practice energy management and create bigger goals for ourselves. To get there, it would be good to understand how we can build a base that enables us to keep going, striving, thriving, and reaching for more, all while juggling families, small businesses, and being what Monica Brown would call "a *bad-ass* boss."

The women I spoke with said that they began their armoring techniques and learning resiliency early in college. About 40% of them attended credited programs like Upward Bound, INROADS, and university-run STEM programs to nurture their budding scientific minds and supporting them in school. However, about 60% of the women I interviewed reported challenges in school at the hands of professors and guidance counselors, including microaggressions and stereotyping. Of those 60% who had negative experiences in their areas of study in college, 10% of the ladies recalled being told they should change their majors from engineering or computer science to something that involved less math.

In the study *I am Committed to Engineering: The Role of Ego Identity in Black Women's Engineering Career Persistence*, researchers found that the nine Black female engineering students who increased their resolve to remain in their engineering programs cited parental support among other factors. They also attributed most of their resolve to the achievement of other undergraduate Black women engineering majors.[69] It is clear from this study that Black women's foundation of resiliency, crafted from familial ties, is built early on, and we take it with us into our careers. Sasha, an IT analyst, says, "I think I had to tap into things that I knew to be true, which is really hard when situations are trying to tell you other things. So, I had to really dig deep and remember. . . how much my parents had sacrificed."

Dr. Tracy Fanara, a New York native with a PhD in Environmental Civil Engineering who is famous for her role on *MythBusters*, knows a great deal about grit and the need to be strong in male-dominated spaces. She told hundreds of STEM-focused students during the online STEM Success conference in November 2020, "If you're in science, you will need resiliency; science is hard. You will need thick skin." The nature of the field of study we have chosen isn't an easy one, and it is filled with lots of failure, so getting used to getting knocked down to get back up comes with the territory. She went

---

[69] Rashunda L. Stitt Richardson, et. al., "'I am Committed to Engineering': The Role of Ego Identity in Black Women's Engineering Career Persistence," *The Journal of Negro Education*, vol. 88, no. 3, 2019, pp. 281-296, doi:10.7709/jnegroeducation.88.3.0281.

on to advise that the trick is to learn to "flip the script" and let what happened to you fuel you instead.

This advice now leads us to weave together a few clues to recenter ourselves regardless of where we are in our careers; it's never too late to reflect and replan. Let's pull out our Let Your Light Shine, Resistance of the Force, and Internal Locus of Control under an Agile thinking exercise. First, back to your light—what is it? What drives you? What is your purpose and passion? What work are you meant to do; who are you meant to serve? For the women I spoke with, their passion and love for science has always been apparent.

Karla, an Information Security Manager, Maxine a VP at a large IT services company, and Kamryn, a senior systems analyst, all agree that it's the love of the industry that keeps them engaged in their work. All three recounted that they are "good at what they do," and the joy of the work outweighs everything else. These women are all fully invested in advancing their careers so nothing will stop them. In fact, Maxine stated, "They can't outwork me. They can't out-design me. They can't out-present me." So, despite the very struggle in the tech game of corporate America, these women are here to stay!

Once you have discovered or rediscovered the light that you give to the world and why you originally embarked on a STEM career, then you must get very clear on your personal values. What line will you never cross, no matter the consequences? Resist the force of going along with what everyone else is doing if it doesn't serve you. Remember that many of the women interviewed throughout this text never played the maid role, such as baking brownies, always taking meeting minutes, or being stuck on the fun committee, because they resisted those roles for themselves from the beginning. They were teaching people how to treat them by setting boundaries that were healthy for them.

They asked questions of themselves like:

What do I consider unethical behavior?

What behavior will I tolerate from others, and how will I handle microaggressions when they occur? For example, will I laugh and then seek to educate? Or will I be very serious and walk the fine line of "likable" versus "comfortable" zones?

What armor will I put on that will be permeable enough to get me through the day without making me cold and hard-hearted?

With a clearly defined purpose and boundaries, we can start to build our goals and objectives. Think about what value you bring to the organization. What are your superpowers? What skills and strengths do you bring that are uniquely yours and help fill a gap the team has that will help them be more successful than they are today?

Next is to remember that now that we have our North Star to guide us in our daily decisions and career journeys, pivots will have to be made as well. Life happens. Life may bring department reorganizations, company acquisitions, and layoffs, all of which can lead to us losing a mentor, advisor, sponsor, or even our own job, therefore greatly impacting our plans. These obstacles do not have to shake us; our circumstances do not define us. Applying Agile thinking to how we will respond when the challenges mount is the best method for building resiliency.

Dr. Carolyn Colleen Bostrack, Founder of Fierce Network, Executive Director at 1LifeFullyLived, and author of *F.I.E.R.C.E, Transform Your Life in the Face of Adversity, 5 Minutes at a Time!*, champions an Agile mindset for learning how to become unmovable and resilient in the face of fear and hardship. Carolyn built up her own reserve of resilience using an Agile mindset as she journeyed from poverty to a powerhouse businesswoman and mother.

Her remarkable story took her from childhood molestation and having to "earn her keep" in the family as her mother reminded her constantly that she was adopted to escaping an abusive marriage becoming a young mother at the age of twenty with no real support system, then becoming a celebrated author, speaker, and coach with a Ph.D. in Organizational Leadership from the Chicago School of Professional Psychology. However, it's the Ph.D. that she got from her life lessons that can help us most. Carolyn teaches us the B.F.F. method:

> **B-Breathe:** Take a deep breath, inhaling through your nose and out from your mouth.
>
> **F-Focus:** What is your immediate goal? What will serve you right now in the moment?
>
> **F-F.I.E.R.C.E. Action:** Select one action that will move you toward your goal.

Small, incremental, iterative steps toward achieving your goal combined with taking time to reflect and holding a personal retrospective on what is working well versus what needs to change are all hallmarks of an Agile mindset that Carolyn teaches to help women become more resilient. This can be put into practice immediately.

We all go from one Zoom call to the next, and half the time we are thinking to ourselves, "Why was I even invited to this meeting?" or, "What do they need from me?" Carolyn advises us to take the three minutes to breathe between calls, center ourselves, and then think about our goals for the meeting. She suggests (just as Sherri Brown Littlejohn coached us in the previous chapter) that our goals can simply be to seek clarification or understanding. She says to "get inquisitive." Then, our "F.I.E.R.C.E. Action" is to ask meaningful questions that focus on gaining clarity from others' comments and help the team summarize the action that will be taken from the discussion.

However, prepare yourself for the various personalities that you will encounter. Oftentimes, we will be in settings with lots of dominant men, or even women who present strong masculine energy, and we should all anticipate that one person with the fragile ego who is wounded and masculine energy from the person that is always looking to throw someone under the bus. In these situations, Carolyn reminds us of how important it is to show up while also remembering to give people space for the power trips too. Following this advice during our interview, she showed me a tiny little palm tree on her desk. She said she keeps it there to remind herself and her clients that "a tree that bends never breaks."

Carolyn also teaches the criticality of creating multiple streams of income for ourselves. This will provide freedom from the golden handcuffs that make us feel trapped by the company. If we keep thinking, *I need this job to pay the mortgage, the tuition, the credit card bill*, then we will be in the position of having to walk on eggshells for fear of losing our jobs. Once we have obtained financial freedom and are on the road to generating wealth for ourselves and the next generation, then the job shifts from something we "have to do" to something we "want to do." Lisa, a cybersecurity consultant, cautions that "if you don't want to be an entrepreneur, that's a personal choice, and [it has] to be evaluated very carefully. Please don't rely on these companies for your sole source of income because. . . they're not loyal."

Pushing the envelope, taking a risky assignment, saying no to always taking meeting minutes, speaking up with confidence despite being ignored, and standing up to the office bully will be much easier when we are operating from a place of liberation and personal power. Finally, Carolyn reminds us that "life isn't getting easier, but our coping is getting better."

If anything, life teaches us that change can happen in an instant, and it can be powerful enough to knock us down if we aren't prepared. Hence the need for us to have a personal mission statement, know our purpose, and cultivate a powerful ally network as well as a strong "sista-girl" support group to keep us steady through life's storms. All these mechanisms can help us manage our energy no matter what we're going through, from being "the only" to dealing with constant microaggressions. "I. . . advise women of color how not to fall into those traps because the energy we spend there fighting those microaggressions are energies we can actually be using to focus on our objectives as opposed to defending who we are. I did a lot of the puffing and the peacocking 'don't you know who I am?' stuff that was irrelevant. It was irrelevant. I understand it now. I did not understand it then," advises Stevie, Senior Research Specialist.

Do you think you need help managing your energy? I sure did, and that is exactly when the universe brought Carla Ogunrinde to me. I had the pleasure of Carla interpreting my personal energy in a profile assessment—not once but twice—and both readings rocked my world! Carla quite literally "read" me in less than 90 minutes, and each time left me in a puddle of tears.

Carla is a certified Energy Leadership Index™ master practitioner in the science of personal energy management, but I'm convinced that it is her spiritual gift. We create energy, emit energy, and sit in energy all day, says Carla.

There are two main types of energy: anabolic and catabolic. Anabolic describes the energy that is constructive, expanding, healing, and growth-oriented. Anabolic energy is used to move us toward positive outcomes and helps with long-term success. Catabolic energy is draining, resisting; it is best used in short-term combative situations in which you need a boost to propel you through the crisis moment. Research shows that the most successful leaders are those with high levels of anabolic energy. Leaders with high levels of catabolic energy tend to have short-term success.[70] Therefore, increasing our anabolic energy is required to help us not just survive but thrive in our careers and in life.

To make sure we continue to thrive and not just survive, I suggest creating a crisis plan and a personal self-care plan. Such a plan, like Dr. Henry-Tett's S.M.A.R.T.L.Y., will help us remember that the importance of managing our energy is necessary—not just for our physical health, but for our mental and emotional well-being too.

My grandmother always tells me, "Angel, you have to stop and smell the roses sometime, honey." I

---

[70] Karen A. Buck and Diana Galer, *Key Factor Revealed for Determining Success in Work and in Life*, Institute for Professional Excellence in Coaching, 2011.

admit that for quite a while I did not heed her advice. I didn't think I could. I didn't believe it was safe for me to take my foot off the gas pedal. However, I was at a conference recently where the speaker helped me shift my perspective on the notion of slowing down and taking time to stop and maybe even take a step back because we need that energy to propel us forward, much like a slingshot. A slingshot move may just be what you need right now, but it will require the courage to slow down, stop, or maybe even take a downward or lateral move.

As all the other coaches I spoke with professed, we need to have a tribe to help support us during our times of need or when we find ourselves in slingshot moments. Tiffany Dufu, author of *Drop the Ball*, encourages us to not just look to mentors, advisors, counselors, and sista-friends, but to discover that help can come from very close to home; in fact, it can come from inside our homes.

If you are married and have a little one or two or three, then you know that the struggle between work life and home life is no joke! The shift to working from home for many tech companies has now blown a hole in the dividing line between work and personal life. Now it's all just blurred together, and the five minutes between meetings to run to the coffee room or chat with a coworker is now filled with making sure your home Wi-Fi is operational and your kid has access to Google Meet to update their assignment. Although, I doubt you really have five minutes since Zoom sessions go back-to-back, and you're Zooming while making lunch and printing homework worksheets.

If you and your spouse are both working from home plus the kids are virtual homeschooling and you find yourself always being the one throwing a load of towels in the wash, preparing meals, and making sure

Michaela and Jesse at least have on clean shirts and their faces washed for their calls, then *Drop the Ball* is a must-read. Tiffany explains how to create a true partnership with our domestic partners, calling them "all in" partners, a partnership that is more equitable and doesn't leave us as women exhausted, resentful, and hard to get along with at home. In summary, primary bread-winning, Type-A, super-intelligent, scientifically minded, results-oriented women are juggling a lot of balls, and Tiffany gives us permission to drop a few.

"All in" spouses, active allies, tribe women, and a personal board of directors are all essential to creating an environment in which we can grow and blossom. Other people pouring in positive affirmations and offering sage advice is great, but if the soil isn't ready to receive the seeds of positivity, they can blow away in the wind. Remember the Bible story the Parable of the Sower? This is when Jesus tells us of a farmer who sows seed indiscriminately. Some seed falls on the path (or by the wayside) with no soil, some on rocky ground with little soil, some on soil which contains thorns, and some on good soil. In the first case, the seed is taken away; in the second and third soils, the seed fails to produce a crop. However, when it falls on good soil, it grows and yields thirty-, sixty-, or hundred-fold. Like any good farmer, we must cultivate our garden, especially the garden of our mind.

Joyce Meyer's bestseller *The Battlefield of the Mind* is one of my favorite resources that I re-read to help keep my mind at a calm and fruitful state. Thorns and weeds can grow in our mind's garden, and Tara Mohr, author of *Playing Big*, describes these thorny, unwanted shoots that prevent good seed from taking root as an "inner critic."

My inner critic's name is Beth. She looks so sweet and cute on the outside. She's average height, average weight, with big eyes as black as marbles and soft, loosely curled hair. She's quiet until she's triggered. Beth can come out swinging when provoked, like Muscle Man Randy Savage, but she doesn't emerge when you think. She's not triggered by some external force seeking to do me harm; she springs into action when I dare to dream. The instant I had the thought to write this book, she came out swinging, saying, "No one will talk to you. Who are you to play junior researcher?" and, "No one cares what you have to say." Beth can be nasty.

 I'm going to venture out and guess that you have a Beth too. She's front and center on your shoulder, whispering to you in the meeting as you try to summon up the courage to share your new idea, and I bet she has shown up as you read the clues throughout this book when you've contemplated putting a few into action. Your inner critic doesn't have to be a woman either. That voice can show up as a male voice, such as an old boss or even your father's voice. Tara characterizes it as the "voice of not me," a voice that tells a woman she is not ready to lead yet, but that she still needs to obtain that degree, get her project management certification, take the class, finish this project, and that all this stuff must be done before she can start doing what she really wants to do. To silence that voice, Tara provides the practical steps needed to shut down your inner critic and instead find your inner mentor.

 My inner mentor's name is Kelly Ann Lee. Kelly Ann's energy emits a positive and welcoming spirit very similar to my grandmother's. She is wise and offers counsel to anyone seeking it. She never wears makeup but instead rocks big colorful earrings and has dreadlocks down her tailbone in a beautiful salt-and-pepper color. She wears the most expensive bohemian broomstick skirts and long, flowing shawls.

If you could see my closet, you would notice that it is a mix of Beth and Kelly Ann right now. Beth likes to wear traditional blue and black slacks and jackets with small earrings and slip-on shoes, which is a contrast to Kelly Ann Lee's relaxed mother-earth elegance. My closet is a manifestation of my inner transition. Beth represents the old external locus of control, a time when I was fearful to speak my pain or be my true self, while Kelly Ann Lee just doesn't give a shit what anyone else thinks.

Carla Ogunrinde told me in our last session together to get down on my knees and give Beth a great big hug and tell her everything is going to be okay, that she doesn't have to be afraid of being hurt, and that she needs to sit down to let Kelly Ann Lee take over. Beth's stubborn, but Kelly Ann is growing stronger each day. Other women have found their inner mentors and allowed them to show up more often as well. "I've grown so much and matured so much as a person into who I am. I praise more to people in my group. I encourage more. I don't take too much to heart," says Monica, head of engineering at a multinational industrial corporation, on her personal growth.

Tara encourages us to "play bigger" with a sizable call to action. She recruits us onto a "transition team" for the world. She explains that our society is in a transition phase. Things aren't quite as racist and misogynistic as they used to be, but we are nowhere near where we need to be regarding the equity and inclusion of all genders, races, ethnicities, and nationalities. So, to be effective at creating a world we want to live in and raise the next generation in, we must become change agents. Change agents can't be held in bondage due to fear; they must trust their inner mentor voices and be able to take large leaps into their purpose, all while performing self-care and cultivating tribes of more powerful women. When Tara Mohr says to "strengthen your ability to be a change agent," she means building resiliency for the work ahead.

Dr. Taryn Marie Stejskal is an international expert on resilience. She is the author of *Flourish or Fold: The Five Practices of Particularly Resilient People*, to be released in 2021. This resilience researcher, educator, speaker, and executive coach offers a bit of scientific basis for all the recommendations and guidance we have heard about building resiliency. Taryn's work focuses on "helping others access resilience in practical ways" based on over a decade of research that has allowed her to discover "the behaviors that support people in developing and sustaining resilience in leadership and in life."

Dr. Taryn's advice is based on the scientific study that includes executive leaders that we would characterize as successful. The science behind resilience is founded in neuroplasticity. She explains that neuroplasticity means that the neurons in our brains, the connections we have, are growing and changing and shifting over time. Maybe you've heard that the neurons that wire together also fire together. We have bundles of neurons that are communicating with one another. When we have particular experiences, our brain, our neurons, and the connections that they have and the way that they're clustered together change, which reflect our experience. Even at a cellular level with our neurons, we are fundamentally and forever changed by every experience that we have. We never go back to the way we were before. We essentially learn to bounce forward.

Through her research, Dr. Taryn has found that our behavior and mindset are in a bi-directional relationship, feeding from each other. Our experiences that we navigated through already, such as sexual harassment, being ignored, and being bullied, do change us, but it is up to us to shape *how* they change us. Shifting from a fixed mindset to one of growth, from survival mode to thriving, and from having a negative experience define who we are to using it to help us grow is required to not become angry and bitter but instead become resilient.

Dr. Taryn has found that there are many ways to define "resilience." She often hears that resilience means the ability to bounce back, managing and adapting to changes with grace, strength, and tenacity, being flexible/agile, and shifting and pivoting when circumstances change. She defines it as the ability "to allow ourselves to effectively address challenges in a way in which we are enhanced by the experience rather than diminished by it." As a result of Taryn's research on resilience from neuroscience to marital relationships, she has found that the most critical and foundational on is practicing vulnerability.

Vulnerability requires a great deal of effort. Learning how to align your external self with your internal self is called "congruence." Achieving this alignment can be a lifelong journey, but every step you take to show more of your true self outwardly is a step toward becoming an authentic and empathetic leader.

We know Black women don't feel emotionally safe to share authentically out of fear. The fears of being stereotyped, discriminated against for telling the truth, or thinking or being different can be root causes of the fear. Being "the only" in male-dominated spaces can also become a place where we sit in a fearful state, but it doesn't have to be once a foundation of resiliency is built. The key to living a resilient life, according to Dr. Taryn, "isn't about being fearless but to simply fear. . . less." We must stop looking at challenges and obstacles as something to "get over," like there are discrete obstacle-based events in our life that happen that we must tackle to overcome and get the victory. Instead, shifting our perspectives to see obstacles as continuous in nature, there will always be something or someone that poses a challenge to the goal that we are trying to obtain; focusing on learning and developing new skills to face the latest trial will ultimately serve us better.

Be mindful of how closely related the words "scared" and "sacred" are in their spelling but also in their meaning. Dr. Taryn encourages us to see the things that scare us the most are really our most sacred development opportunities. In summary, she states that "we have to learn to coexist with challenges and get comfortable with challenges being never-ending." Quoting Pema Chodron, she said, "Nothing ever goes away until it teaches us what we need to know."

As so many of our coaches have recommended, Dr. Taryn reminds us to build our tribes or villages. That means leveraging a strong support network to create confidence, increase self-esteem, and practice being truthful by sharing thoughts, opinions, and ideas. Strengthening yourself is akin to building muscle, so the same exercises and repetition done for a physical workout is needed to allow ourselves to flourish internally as well. We need to access the resources, the knowledge, the wisdom, and the community support often to effectively address our current challenges. We can use our villages/tribes to nourish the soil in which we will grow as leaders. Dr. Taryn's final advice is to "invest in yourself!"

Hopefully all this talk and expert advice on resiliency has convinced you that it is the key to help you achieve your dreams, goals, and live out your personal mission. I will push this thought just a bit further and ask you to consider that you not only need resilience but extreme resilience.

Resiliency Building

Dr. Srikumar Rao, TEDx speaker and author of *Are YOU Ready to Succeed?* and *Happiness at Work,* has helped thousands of entrepreneurs and executives worldwide achieve quantum breakthroughs in their personal and professional lives by using the thought of extreme resiliency. He suggests that eliminating stress, live your biggest dreams and showing up in the world exactly like you envision your inner mentor requires extreme resiliency, which will help you make some quantum leaps in your life. Creating massive resiliency will require substantial changes to your thinking patterns. Your view of life dramatically impacts your ability to jump and leap.

During his final keynote address during 100 Leaders Live, an online conference in October 2020, Rao offered the example of a bricklayer, whose story was near and dear to my heart as my grandfather was a retired bricklayer. Rao recited a tale of an architect in ancient England who was searching for a bricklayer to build a cathedral. When the first bricklayer was asked, he responded that his job was to "just break rocks," then the second bricklayer said that he was helping to build a wall, and the third bricklayer realized that the back-breaking labor was worth it for the opportunity to learn how to build a cathedral. Twenty years later, the guy who was "just breaking rocks" was dead. The guy who was helping build a wall was living a life of misery. But the guy who was helping build a cathedral was on his way to building his first cathedral. Rao commented:

> "That is a choice that every one of us has every day. We can get up in the morning and we can break rocks, or we can get up in the morning and help build a cathedral. I cannot define for you the cathedral that you are building—you are the only person who can do that. But I can

tell you that unless you define for yourself the cathedral that you're building and anchor yourself in it, you're going to live a mediocre existence."[71]

Women scientists have always seen themselves as "cathedral builders." The women I interviewed for this book took the time to reflect on what they would say to the next generation about what it takes to "make it" in our industry while working in corporate America, the same as the minority women scientists who participated in the 1975 report *The Double Bind: The Price of Being a Minority Woman in Science*:

> Our mission at this meeting was clear. We wanted to find out how and why we had made it and others had been left behind; how our sisters had handled personal and societal problems from childhood until the present. We discovered, despite differences in minority cultures, that we as women scientists (a) read at an early age; (b) had a strong sense of self; (c) were always aware of our ethnic status; (d) remembered the encouragement of a particular teacher or friend; (e) were rarely ambivalent about school and further education; (f) were disciplined to study and (g) were aware of our sex in a positive way.[72]

---

[71] Srikumar Rao, "Resilience is Wonderful. Extreme Resilience is Far Better!," *100 Leaders Live Summit*, Oct. 15, 2020.

[72] Shirley Mahaley Malcom, et. al., *The Double Bind: The Price of Being a Minority Woman in Science*, Report no. 76-R-3, American Association for the Advancement of Science, 1976.

Fast forward 45 years later, and the women I interviewed had several lessons to pass on to the next group of emerging Black women in tech.

Frontline Tales

> You have to take onus, too, because it may be the way that you're communicating. And that was something that I am learning as well, is that it may not be that I'm being intentionally ignored; I need to also evaluate how I'm communicating. But, even still, justifiably, you could possibly be literally being ignored. And one thing that I learned—one of my teachers said in one of our classes is that the human mind—you need to repeat things three times in order for it to stick.
>
> —Kamryn, Senior Systems Analyst of Energy Company

> Yeah, because like I said, it was really my mission. I was like, "I cannot leave this program and not put somebody in this program behind me." But I remember what I told her, and what I would tell most young women that I meet or anyone that I do meet is that you deserve to be here, number one. Don't internalize anything anyone else is trying to say. Don't think that you're not good enough. Don't think that you're here because you're Black. You're here because you're smart. You're here because you're capable. And this is your time to shine and to do the work.
>
> I'm very spiritual, too, so I was just like, "I don't think God would have placed me—this was not

on my plan, my path. So, I don't think God would place me here if it wasn't something that I could do." So, I really had to dig deep and channel into that and find out what I knew to be true, which is that one, I deserve to be here; that's why I'm here. Not because I'm Black, not because I'm a female, but because I know how to do the work. And I happen to be Black and to be a female.

—Sasha, IT Analyst at Fortune 500 Computer Software Company

Understand that you are NOT crazy. The little attacks, attitudes, hidden biased behavior is a full attack. Breathe, and remember you are enough, you are worthy, you are smart, you are right. Stay calm, and create a few strategies, find options.

—Terry, Director of Strategic Transformation of Automation Company

Even though we were in IT and, as you know, business casual, I would wear suits. And then, at a point, I started to mostly wear pantsuits because I didn't want you to see me as the female. I want you to see me as, "Here's the professional." I would always wear heels because I wanted to be tallest. And those were ways that, for me, helped to make me feel as though it would take away some of their ammunition. If you aren't physically looking down on me, then hopefully you won't mentally be looking down on me. I would go into the room first and raise my chair all the way to the top in the conference room.

—Karla, Information Security Manager

Look back at all your successes and wins often so that you can remember who you are, remember what you can do, and then you say, "If I did that, I can do this. If I was able to get through that, I can get through this."

—Tracey, Director of IT at
Electric Manufacturing Company

You have to develop community, your family, your church, because that is what will sustain you. If you put everything into the job, basically, they can take the bottom right from under you, and you have nothing. So, if you invest in your community, invest in your family, invest in your neighborhood, invest in your church, you have something to hold on to.

—Charmaine, Senior Government Aeronautical
Staff Member

So, I know that I came out on the good, that my life, my career has been more blessed, more than it ever would have been there. I'm happier than I think I would have ever been. . . So, I came out good, which is why I'm [like], "Hey, don't sweat it. It all worked out how it was supposed to work out."

—Mary, Research Patent Attorney

Regardless of what this person says to you or what your financial situation is, God will always provide and take care of you. But you don't ever want to look back on a situation and say you wish you could have, should have, would have. No.

Always be true to yourself and respect yourself. Okay? And respect others, no matter what, because at the end of the day you want to treat people the way that you want to be treated, period. And to return evil for evil—you have to just put people in God's hands and do right by yourself, do right by others, and know that it'll all work out, period. Never compromise your integrity.

—Dominique, IT Validation Specialist

There's absolutely a rearing culturally of us of "what doesn't kill you makes you stronger." And I think our guardians failed to qualify that statement by saying that only works with God. I don't put my faith in a man. A man can be used by any and everything. But God—what doesn't kill me from God will make me stronger because that's biblical. But when it comes to men, they will try to kill you, and they might succeed. So, don't leave your hands and your faith in a man.

—Diane, Senior Business Consultant at Big Five Consulting Firm

I subscribe to an online Bible Study app called FirstFive. The mission is simple: give God the first five minutes of your day, every day. The compact bible study lesson is extremely helpful and my favorite part of my morning routine. As I write this conclusion on resiliency, I would be remiss if I didn't tell you how I get through my days.

I'm no longer a junior program manager struggling for credibility or an analyst who feels she has to jump through hoops and can't push back for fear of retribution. My confidence has grown, I've been through at least eight reorganizations, two job eliminations, and have waded the waters of the after-effects of two companies merging . . . AND I have no idea what will come next . . . AND I'm okay with not knowing.

I do have a budding active ally group of men supporting me, a hilarious sista-girl tribe of sorors, and a phenomenal amount of good karma in my local tech community. I don't believe in burning a bridge or not helping someone every chance I get, but all those people have the potential to fail me when I need them most. They are human and well-intentioned but fallible still. It's my growing faith in my Savior, Jesus Christ, and knowing that He will sustain me, He will see me through, and He will bless me at all times through the storm, the rain, the reorganization, and the layoff. I have no doubt that my God loves me, and His love is my foundation, my North Star, and my protection. So, when all the clues we tried fail, when we've done our best to improve our station in life and make the job work, but we get stuck call on the name of Jesus, and he will do what he promised to do, which is answer your prayer, so long as it is according to His will.

And now, honestly the not-so-much-fun part begins, and that is the waiting. The waiting can be brutal, but it doesn't have to be. While you wait for your next move, promotion, recruiter to call, supervisor to leave. . . whatever you are praying and waiting for, you can still be fruitful. As Sarah E. Frazer, FirstFive writer, notes, the situations in our lives can be summarized in three acts.

Act One: God promises.

Act Two: Pray and wait.

Act Three: God answers.

Although FirstFive Bible lessons are edifying, the comments at the end are where I derive the most value. Comments at the close of the lesson come from women all around the globe who access the app, women just like me—mothers, corporate mavens, wives, daughters, granddaughters—all remarking on how the daily lesson impacts their lives. This comment from a FirstFive subscriber challenged me, and hopefully will challenge you as well. Prayerfully, it stimulates you to set your unmovable, unshakable, solid-as-a-rock foundation on He who longs to be your everything, because the work you have to do is much, much bigger than the job you have at that company.

> "As I make plans, career plans for the future, I find myself saying 'I' way too much. 'I' pitched this new position, 'I' will be so disappointed if it doesn't happen. Instead of relying on God to reveal where he wants me to be. Asking God to put me where I cannot just advance my career but advance the gospel. When I come at it from that angle, and leave it up to God, then I don't need to be anxious or stressed about it because I can trust that it's all in God's hands for the purpose of His glory."

And we know that all things work together for good to those who love God, to those who are the called according to His purpose.

—Romans 8:28 (NKJV)

Epilogue

I started with a premise that the boys in Silicon Valley and the guys sitting at every C-Suite table in corporate America were confused. I pictured them scratching their heads, wondering where all their money went. You know, the missing hundreds of thousands—no I mean millions, oh sorry, BILLIONS of dollars—invested in diversity and inclusion programs over the last decade. According to Pamela Newkirk, in her book *Diversity Inc.: The Failed Promise of a Billion-Dollar Business*, the diversity and inclusion space has become commodified and essentially exploited. Fake programs led by charlatan diversity consultants are not only not improving corporate America, but they are making it worse. The boys in the jeans and button-down sweaters with flip flops must be irate over the loss of their money!

However, after a period of year-long research and hearing African American women's tales from the

Epilogue

underbelly of corporate America, I have changed my mind. I think they know exactly where their money is going. They look at it as lawsuit prevention and marketing campaigns. The portion of their budget reserved for diversity and inclusion work will not only secure D&I grants from federal and state government programs but help protect them from lawsuits if a formal complaint is lodged, like the one my pregnant self submitted naively when I was yelled at like a toddler by a man with fewer degrees and less leadership ability. Organizations have a built-in self-protection, get-out-of-jail-free card when they administer diversity training as the resolution. Also, the bar is pretty damn low for these companies to make improvements in their hiring and recruiting numbers.

If a company has 3% African American or Latinx working for their company one year, but by the subsequent year they have 6%, call it success! No attention at all is given toward retention, sponsorship, or a fast-track path toward upward mobility like the internal leadership programs designed for their young, White, and highly networked counterparts. You hear all day long about mentorship programs for women and minorities, but I have yet to encounter a sponsorship program—they exist, just not as prevalently. Despite the alphabet soup of letters behind my name, I'm left to my own devices to grind it out and figure out how to "gain more experience."

Don't take my word for it or the word of the ladies whose experiences you've read here... a quick Google search will yield the bleak and troubling numbers revealing that not only is unconscious bias training, diversity training, and sexual harassment training not working, but our number of representations at mid-level manager and above are going backward! Of C-Suite leaders today, 21% are women and just 1% are

Final Note to My Emergent Active Allies

Black women. Black women hold only 1.6% of VP roles and 1.4% of C-Suite executive roles.[73]

So, now my focus shifts. Instead of trying to convince the top echelons of corporate America of what is really going on in their meeting rooms, conference halls, and one on one meeting rooms, I turn to the next generation and their allies.

First, to the young ladies who survived a STEM program in Computer Science, Information Systems, Cybersecurity, Applied Mathematics, or Engineering... CONGRATULATIONS! I am so proud of you, happy for you, and I applaud your tenacity. I am just here to tell you that after two decades of surviving in the trenches, now is not the time to rest. Armor up, dig in, hit the virtual front door of your new job or internship looking to build your network, find a sponsor, and work your tail off. Don't make the mistake of putting your head down, working hard, and hoping someone will notice your efforts. Follow Harvey J. Coleman and "manage your PIE'" from the very first day.

To my allies—those of you who coach, mentor, advise, sponsor, and have a willing heart to see equality—I applaud you as well. I thank you for tapping into your humanity and seeing that when a policy isn't right, a candidate pool isn't diverse enough, or when an African American woman is "the only" on your team, that you say something. (Thanks, Dave.) Be an active ally by speaking up and asking the tough questions, bringing attention to bias practices, and calling out coworkers when inappropriate jokes are made and unfair treatment is doled out. No matter how minor or micro it may seem, you now know how the weight of

---

[73] *The State of Black Women in Corporate America*, Lean In, 2020, Lean In.org/research/state-of-black-women-in-corporate-america/introduction.

the "little cuts" build, and you know now that your silence is deafening.

To those with a seat at the table, if you have hiring and recruiting responsibilities and have the power and authority to set the culture in your company, use your knowledge for good. Create sponsorship (not mentorship) programs for women and minorities, research what the Top 50 companies did with D&I work to make real and sustainable change, and finally, hold your peers accountable for "putting their money where their mouths are." Most importantly, do not let your organization's D&I initiatives stop at just entry-level positions. You must ensure that minority women—not just White women—are hired at every level of management. We are ready now—you don't have to wait to "grow the pipeline" of diverse talent. We are here and ready to lead your organizations through the next crisis and beyond.

Resources

Toolkit for African American Women

Empowering Yourself: The Organizational Game Revealed, 2nd Edition
Harvey Coleman

Expect to Win: 10 Proven Strategies for Thriving in the Workplace
Carla A. Harris

Forget a Mentor, Find a Sponsor: The New Way to Fast-Track Your Career
Sylvia Ann Hewlett

What Got You Here Won't Get You There
Marshall Goldsmith & Mark Reiter

Resources

How Women Rise: Break the 12 Habits Holding You Back from Your Next Raise, Promotion, or Job
Sally Helgesen

F.I.E.R.C.E.: Transform Your Life in the Face of Adversity, 5 Minutes at a Time!
Carolyn Colleen

Toolkit for Allies

Lead Like an Ally: A Journey Through Corporate America with Proven Strategies to Facilitate Inclusion
Julie Kratz

Network Beyond Bias
Amy C. Waninger

Hire Beyond Bias
Amy C. Waninger

Emotional Intelligence: Why It Can Matter More Than IQ
Daniel Goleman

Acknowledgements

A sincere thank you to all the women who took the time to speak to me. Our sessions were a mix of two old friends who haven't spoken in a long time (despite it being our first encounter for many of you) and a counseling session. I appreciate that discussing your personal stories, no matter how long ago the incidents occurred, was not easy. My prayer is that each person reading this book not only empathizes but is moved to action because of your bravery.

Thank you to my family for enduring the nights and weekends of research and writing. I want my sacrifice to be meaningful to you as well. This is my demonstration of how sharing your life lessons can impact someone else. If my heartache can be used to ease another's pain or better yet prevent their distress, then it was worth it.

To my EMERGE SiSTAR group, 2020 Vision—we went through a global pandemic and came out for the better! Thank you for your prayers, encouraging words, and the most organized food train that I have ever seen! My family suffered quite a few trials and tribulations in a short 12-month period, and you were right there with me.

To my close-line sisters and close Sorors of Zeta Phi Beta Sorority. You have blazed life's trails with me and supported me every step of the way. I love you! I've known all five of you for almost 20 years, and I look forward to the next 20. They ain't ready for us!

Thank you to my Indy-based female Agile tribe of the most energetic, uber-smart, take-no-prisoners, grab-life-by-the-balls ladies I have ever had the privilege of meeting! You ladies push me every day to wake up and do great things for my teams and my company. Your true support and daily encouraging messages are a bright spot, and if we ever can organize our work and family schedules to take our much-talked about "Tribe Trip," I feel sorry for the city we select. They ain't ready for us!

I am so appreciative of each and every coach, wellness expert, and diversity and inclusion professional that took time from their practice to speak with me. I am floored that every single one of you I reached out to for guidance on what to tell women to do to thrive in the face of so many obstacles eagerly answered my call, LinkedIn request, or meeting invite without hesitation. Your dedication to creating inclusive spaces is remarkable! I am privileged to know and work with so many people truly making this world better for others. This is "wonder if it is worth it" work, and while this journey of unmuting women has proven to be many things, it has never been lonely. Thank you for walking this path this us.

I just love my allies. It takes a special person to cross the color divide in this county and lend their political or social power to help someone else. I'll never forget when Alan yelled at Tom for not allowing me to go home early during my internship when I was so sick I could barely type. Thanks, Alan, for standing up to help a girl while she was down! Thank you, Bradley, for spending your evenings away from your wife and kids to teach me MS Project as well as how to lead team meetings. Not sure where I'd be without my career-long co-conspirator Dave, who has the superhuman ability to be advisor, mentor, AND sponsor all in one and wear the biggest smile on his face while doing it. It must have been that special water at the water fountain, huh, Dave? Thank you to JoAnn for believing me, pushing me to excel, and sponsoring me to get my PMP—just know that your management style and leadership mantras have become a blueprint that I follow to this very day! To Mary, both Scotts, Julie, Michelle, Trish, Marti, Shannon, and Don, who have sponsored me along the way and are in my corner. There are no words. Your ability to see the potential in others and match them to opportunities is exceptional.

Everyone's participation has been sanctioned by God. The idea of talking to other women to ask them about their stories all the way to writing the final chapter has been a step-by-step unveiling of God's will. Tim, Tonya, Eric, and Kellie, your influence in my life was clearly God-ordained and propelled me to keep up this work. I can say for certain that God sees and hears each of us, even at our lowest point—especially at our lowest point-—and has not forgotten us. Keep fighting this fight for equity and inclusion. Dent-making isn't easy, and although you will get tired, never give up!

Dents in the Ceiling

Bibliography

"50 Ways to Fight Bias, a Bias Program to Support Women at Work." *Lean In*, https://Lean In.org/gender-bias-cards/grid/card/set-4/14. Accessed 17 Mar. 2021.

"Advisor." *Dictionary by Merriam-Webster: America's Most-Trusted Online Dictionary*. Dictionary by Merriam-Webster: America's Most-Trusted Online Dictionary, https://www.merriam-webster.com/. Accessed 17 Mar. 2021.

"Ally." *Dictionary by Merriam-Webster: America's Most-Trusted Online Dictionary*. Dictionary by Merriam-Webster: America's Most-Trusted Online Dictionary, https://www.merriam-webster.com/. Accessed 17 Mar. 2021.

Bibliography

Barlow, Jameta Nicole. "Black women, the forgotten survivors of sexual assault." *American Psychological Association,* Feb. 2020, https://www.apa.org/pi/about/newsletter/2020/02/black-women-sexual-assault.

"Black Survivors and Sexual Trauma - TIME'S UP Foundation." *TIME'S UP Foundation,* https://www.facebook.com/TimesUpNowOfficial/, 20 May 2020, https://timesupfoundation.org/black-survivors-and-sexual-trauma/.

Brandon, Rick, and Marty Seldman. *Survival of the Savvy.* Simon and Schuster, 2004.

Buck, Karen, and Diana Galer. "Key Factor Revealed for Determining Success in Work and in Life." *Institute for Professional Excellence in Coaching,* 2011, https://alacartecoaching.com/download/Key-Factor-For-Success.pdf.

Capehart, Jonathan. "Starbucks COO Rosalind Brewer on Race and Leadership." *Washington Post,* The Washington Post, 3 Dec. 2019, https://www.washingtonpost.com/opinions/2019/12/03/starbucks-coo-rosalind-brewer-race-leadership/.

Carpenter, Julia. "The 'emotional Tax' Afflicting Women of Color at Work." *CNNMoney,* 5 Mar. 2018, https://money.cnn.com/2018/03/05/pf/emotional-tax-women-of-color-at-work/index.html.

"Black Female Entrepreneurs Look for Ways to Improve Diversity in Silicon Valley." *YouTube,* uploaded

Bibliography

by CBSN, 29 Jan. 2019, https://www.youtube.com/watch?v=G-GDwC6KN6s.

Clark, Rodney, et al. "Large Arterial Elasticity Varies as a Function of Gender and Racism-Related Vigilance in Black Youth." *Journal of Adolescent Health*, no. 4, Elsevier BV, Oct. 2006, pp. 562–69. Crossref, doi:10.1016/j.jadohealth.2006.02.012.

Comaford, Christine. "How to Stop Workplace Bullies In Their Tracks." *Forbes*, Forbes, 12 Mar. 2014, https://www.forbes.com/sites/christinecomaford/2014/03/12/bust-workplace-bullies-and-clear-conflict-in-3-essential-steps/?sh=74247fb57912.

Dickens, Danielle D., and Ernest L. Chavez. "Navigating the Workplace: The Costs and Benefits of Shifting Identities at Work among Early Career U.S. Black Women." *Sex Roles*, no. 11–12, Springer Science and Business Media LLC, Sept. 2017, pp. 760–74. Crossref, doi:10.1007/s11199-017-0844-x.

Dufu, Tiffany. *Drop the Ball*. Flatiron Books, 2017, pp. 205.

Ellis, Judy T. "Sexual Harassment and Race: A Legal Analysis of Discrimination." *Notre Dame Journal of Legislation*, vol 8. no. 1, 1981, pp. 41-42.

Frye, Jocelyn. "Racism and Sexism Combine to Shortchange Working Black Women." *Center for American Progress*, 22 Aug. 2019, https://www.americanprogress.org/issues/women/news/2019/08/22/473775/racism-sexism-combine-shortchange-working-black-women/.

Bibliography

Gillespie, Marcia A. "We Speak in Tongues." *Ms.* Jan.-Feb., 1992, pp. 41-42.

Goyder, Caroline. "The Surprising Secret to Speaking with Confidence." *YouTube*, uploaded by TEDx Talks, 25 Nov. 2014, https://www.youtube.com/watch?v=a2MR5XbJtXU.

Green, Tai, and Sylvia Ann Hewlett. "The Center for Talent Innovation Is Now Coqual." *The Center for Talent Innovation Is Now Coqual*, http://www.talentinnovation.org/publication.cfm?publication=1460. Accessed 17 Mar. 2021.

Hall, J. Camille, et al. "Black Women Talk About Workplace Stress and How They Cope." *Journal of Black Studies*, no. 2, SAGE Publications, June 2011, pp. 207–26. Crossref, doi:10.1177/0021934711413272.

Henry, Angel. "Work Allies: Black and White Make Green." *LinkedIn*, 9 Jan. 2020, https://www.linkedin.com/pulse/work-allies-black-white-make-green-henry-mlis-mba-pmp-csm-popm/.

"History & Mission." *INROADS*, https://inroads.org/about-inroads/history-mission/. Accessed 17 Mar. 2021.

Hollis, Leah P. "Bullied Out of Position: Black Women's Complex Intersectionality, Workplace Bullying, and Resulting Career Disruption." *Journal of Black Sexuality and Relationships*, no. 3, Project Muse, 2018, pp. 73–89. Crossref, doi:10.1353/bsr.2018.0004.

Janssen, Markus, et. al., "Qualitative Content Analysis—and Beyond?," *Forum: Qualitative Social Research*, vol. 18, no. 2, 2017, doi.org/10.17169/fqs-18.2.2812.

Jefferson, Erika. "Where Are the Black Women in STEM Leadership?." *Scientific American Blog Network*, Scientific American, https://blogs.scientificamerican.com/voices/where-are-the-black-women-in-stem-leadership/. Accessed 17 Mar. 2021.

Jones, Charisse, and Kumea Shorter-Gooden. *Shifting*. Harper Perennial, 2004.

Kleeman, Sophie. "9 Inspiring Women Leaders in Tech Share Career Advice Everyone Needs to Hear." *Mic*, Mic, 21 July 2015, https://www.mic.com/articles/122556/career-advice-quotes-from-women-leaders-in-tech.

Lorde, Audre. *Sister Outsider*. Penguin Classics, 2020.

Malcom, Shirley M., et. al. *The Double Bind: The Price of Being a Minority Woman in Science. American Association for the Advancement of Science.* 1976. http://web.mit.edu/cortiz/www/Diversity/1975-DoubleBind.pdf.

McCluskey, Audrey Thomas. "Setting the Standard: Mary Church Terrell's Last Campaign for Social Justice." *The Black Scholar*, no. 2–3, Informa UK Limited, Mar. 1999, pp. 47–53. Crossref, doi:10.1080/00064246.1999.11430962.

McGee, Ebony O., and Lydia Bentley. "The Troubled Success of Black Women in STEM." *Cognition and Instruction*, no. 4, Informa UK Limited, Aug. 2017, pp. 265–89. Crossref, doi:10.1080/07370008.2017.1355211.

"Mentor." *Dictionary by Merriam-Webster: America's Most-Trusted Online Dictionary*. Dictionary by Merriam-Webster: America's Most-Trusted Online Dictionary, https://www.merriam-webster.com/. Accessed 17 Mar. 2021.

Meyerson, Debra. "Radical Change, the Quiet Way." *Harvard Business Review*, 1 Oct. 2001, https://hbr.org/2001/10/radical-change-the-quiet-way.

"Microaggression.", *Oxford English Dictionary*, http://oxfordlearnersdictionaries.com/us/definition/english/microaggression. Accessed 17 Mar. 2021.

Newkirk, Pamela. *Diversity, Inc.* Hachette UK, 2019.

Nordquist, Richard. "Learn the Function of Code Switching as a Linguistic Term." *ThoughtCo*, 2019, https://www.thoughtco.com/code-switching-language-1689858.

Ontiveros, Maria. "Three Perspectives on Workplace Harassment of Women of Color." *Golden Gate University Law Review*, vol. 23. no. 3/4, 1993, https://digitalcommons.law.ggu.edu/cgi/viewcontent.cgi?article=1600&context=ggulrev.

Page, Kira. "The 'Problem' Woman of Colour in NonProfit Organizations." *COCo*, 8 Mar. 2018, https://coco-net.org/problem-woman-colour-nonprofit-organizations/.

Bibliography

Petrakis, Melissa. *Re: What is the difference between narrative analysis and thematic analysis? Is thematic analysis an approach of narrative analysis? Research Gate*, 2017, https://www.researchgate.net/post/What-is-the-difference-between-narrative-analysis-and-thematic-analysis-Is-thematic-analysis-an-approach-of-narrative-analysis/58ca7c3ef7b67ef27153937f/citation/download.

Purdie-Vaughns, Valerie, and Richard P. Eibach. "Intersectional Invisibility: The Distinctive Advantages and Disadvantages of Multiple Subordinate-Group Identities." *Sex Roles*, no. 5–6, Springer Science and Business Media LLC, Apr. 2008, pp. 377–91. Crossref, doi:10.1007/s11199-008-9424-4.

Reader, Ruth. "How Silicon Valley's Diversity Problem Created A New Industry." *Fast Company*, Fast Company, 5 July 2016, https://www.fastcompany.com/3061457/how-silicon-valleys-diversity-problem-created-a-new-industry.

Resendez, Miriam G. "The Stigmatizing Effects of Affirmative Action: An Examination of Moderating Variables." *Journal of Applied Social Psychology*, no. 1, Wiley, Jan. 2002, pp. 185–206. Crossref, doi:10.1111/j.1559-1816.2002.tb01426.x.

Robinson, Bryan. "New Study Says Workplace Bullying On Rise: What You Can Do During National Bullying Prevention Month." *Forbes*, Forbes, 11 Oct. 2019, https://www.forbes.com/sites/

bryanrobinson/2019/10/11/new-study-says-workplace-bullying-on-rise-what-can-you-do-during-national-bullying-prevention-month/#5fa852842a0d.

Rudman, Laurie A., et al. "Reactions to Vanguards." *Advances in Experimental Social Psychology*, Burlington Academic Press, 2012, pp. 167–227, http://dx.doi.org/10.1016/b978-0-12-394286-9.00004-4.

Santos-Longhurst, Adrienne. "High Cortisol Symptoms: What Do They Mean?" *Healthline*, Healthline Media, 31 Aug. 2018, https://www.healthline.com/health/high-cortisol-symptoms.

Solórzano, Daniel, et al. "Critical Race Theory, Racial Microaggressions, and Campus Racial Climate: The Experiences of African American College Students." *The Journal of Negro Education*, vol. 69, no. 1/2, Journal of Negro Education, pp. 60–73, doi:10.2307/2696265. Accessed 17 Mar. 2021.

"Sponsor." *Dictionary by Merriam-Webster: America's Most-Trusted Online Dictionary*. Dictionary by Merriam-Webster: America's Most-Trusted Online Dictionary, https://www.merriam-webster.com/. Accessed 17 Mar. 2021.

Stallings, Erika. "When Black Women Go From Office Pet to Office Threat." *ZORA*, Medium,. 16 Jan. 2020, https://zora.medium.com/when-black-women-go-from-office-pet-to-office-threat-83bde710332e.

Bibliography

"Statistics of Black Women and Sexual Assault." *The National Center on Violence Against Women in the Black Community*, 2018, https://ujimacommunity.org/wp-content/uploads/2018/12/Ujima-Womens-Violence-Stats-v7.4-1.pdf.

Stitt, Rashunda L., and Alison Happel-Parkins. "'Sounds Like Something a White Man Should Be Doing': The Shared Experiences of Black Women Engineering Students." *The Journal of Negro Education*, no. 1, Journal of Negro Education, 2019, p. 62. Crossref, doi:10.7709/jnegroeducation.88.1.0062.

Stitt Richardson, Rashunda L., et al. "'I Am Committed to Engineering': The Role of Ego Identity in Black Women's Engineering Career Persistence." *The Journal of Negro Education*, no. 3, Journal of Negro Education, 2020, p. 281. Crossref, doi:10.7709/jnegroeducation.88.3.0281.

"The State of Black Women in Corporate America." *Lean In*, https://Lean In.org/research/state-of-black-women-in-corporate-america/introduction. Accessed 17 Mar. 2021.

Travis, Dnika J. and Jennifer Thorpe-Moscon. "Day-to-Day Experiences of Emotional Tax Among Women and Men of Color in the Workplace." *Catalyst*, 2019, https://www.catalyst.org/wp-content/uploads/2019/02/emotionaltax.pdf.

Wabuke, Hope. "'Caste' Argues Its Most Violent Manifestation Is In Treatment of Black Americans." *NPR*, Aug. 10, 2020, https://www.npr.org/2020/08/10/900274938/caste-argues-its-most-violent-manifestation-is-in-treatment-of-black-americans.

"Women Business Leaders: Global Statistics." *Catalyst*, https://www.facebook.com/catalystinc/, https://www.catalyst.org/research/women-in-management/. Accessed 17 Mar. 2021.

"Women in the Workplace 2020." *Lean In*, https://LeanIn.org/women-in-the-workplace-report-2020. Accessed 17 Mar. 2021.

"Women, Minorities, and Persons with Disabilities in Science and Engineering: 2019." *National Center for Science and Engineering Statistics*, National Science Foundation, https://ncses.nsf.gov/pubs/nsf19304/. Accessed 17 Mar. 2021.

Workplace Bullying Institute. https://workplacebullying.org/. Accessed 27 Feb. 2021.

About the Author

Angel G. Henry has a passion for diversity in the field of IT. Her knowledge of why women and minorities are oftentimes missing from the C-Suite provides awareness for IT leaders to drive change. Angel has over 20 years of IT experience, primarily in the pharmaceutical and healthcare industries, and over 15 years in the project management discipline. She has become a recognized thought-leader on the topic of the Agile mindset which fosters an environment of innovation and productivity.

Angel is also a sought-after trainer and speaker as well as an adjunct instructor in the Indianapolis area helping students realize their full potential in the field of IT. She is a member of several professional and community organizations, including BDPA, Information Technology Senior Management Forum (ITSMF)-Executive, a member of several Agile communities, and Zeta Phi Beta Sorority, Inc.

Dents in the Ceiling is Angel's first book.

To watch an interview with Angel scan below:

CPSIA information can be obtained
at www.ICGtesting.com
Printed in the USA
JSHW021102050621
15581JS00001B/3